A Color Atlas of

DANGEROUS MARINE ANIMALS

A Color Atlas of DANGEROUS MARINE ANIMALS

BRUCE W. HALSTEAD, MD
Director, World Life Research Institute
Colton, California, USA

PAUL S. AUERBACH, MD, MSM
Associate Professor of Surgery and Medicine
Director, Emergency Services
Vanderbilt University Medical Center
Nashville, Tennessee, USA

DORMAN R. CAMPBELL, MA
Director of Education, World Life Research Institute
Colton, California, USA

CRC Press, Inc.
Boca Raton, Florida

THIS BOOK IS DEDICATED TO
Terri Lee Halstead; the Auerbach children, Brian and Lauren; and the Campbell children, John-Paul, Jeremy, Briony and Bruce.

Distributed in Continental North America, Hawaii, Puerto Rico by CRC Press Inc., 2000 Corporate Blvd, NW. Boca Raton, Florida 33431, USA, by arrangement with Wolfe Publishing.

Printed by W. S. Cowell, Ipswich, England.

ISBN 0-8493-7139-2

CONTENTS

PREFACE

Underwater activity has increased in recent years at an explosive rate. Thanks to the ready availability of the aqualung and modern jet travel, far away places and tropical seas have become the diving domain of the ordinary sport diver. Furthermore, the continual search for new energy sources has made the exploration of the ocean floor a high priority. Drilling platforms and a sophisticated submarine technology have taken divers to ever increasing depths and to some of the most adverse oceanic regions.

Marine archeology has opened new vistas to the underwater cartographer. Painstaking records of ancient shipwrecks on the ocean bottom demand hours of hard work under difficult circumstances. Photographers have taken the underwater plunge and are now providing the public with a kaleidoscope of colorful creatures previously seldom seen by even the most knowledgeable of marine scientists.

Modern medicine is also in the process of getting its feet wet, beginning to explore the underwater world in quest of new biodynamic substances having pharmaceutical potential. The aqualung has now been added to the scientific tools of pharmacologists, pharmacognosists and chemists. Noxious poisonous creatures may ultimately be found to be purveyors of life-saving drugs momentarily disguised as dangerous marine organisms.

Marine biologists no longer limit themselves to stuffy museums and pickled blanched out specimens, but are studying living creatures in vivid color – thanks to the advent of the aqualung. And there are a host of others such as oceanographers, television personalities, writers, adventurers, ad infinitum, all of whom have succumbed to the lure of Cousteau's "silent world".

The magnificent underwater world, however, is not without its problems. In this book we have tried to address some of the challenges of underwater wilderness medicine, presenting both the problems and the solutions when known. A critical review of this matter clearly suggests that we seem to have a disproportionate number of problems as yet unresolved when compared to the solutions that have been found. Nevertheless, essential preventive data are abundantly provided in this book which can save the reader a vast amount of physical discomfort if properly heeded in a timely manner.

It is only fitting and proper to acknowledge the fact that the information provided in the pages of this book is the essence of a distillate of scientific data resulting from travels in more than 160 countries and an enormous amount of research over the past 40 years. This work was sponsored by numerous gifts, grants and contracts from individuals, private organizations and governmental agencies, a partial listing of which include the following: School of Aerospace Medicine, Air Force Office of Scientific Research, and Science Division, Directorate of Science and Technology, Department of the U.S. Air Force; Office of Naval Research, Bureau of Medicine and Surgery, Department of the U.S. Navy; Chemical and Radiological Laboratories, Office of the Surgeon General, Department of the Navy; National Institutes of Health, U.S. Public Health Service; National Science Foundation; Armed Forces Institute of Pathology; U.S. Coast Guard; Pacific Oceanic Fisheries Investigations, U.S. Fish and Wildlife Service; Van Camp Laboratories; Food and Agriculture Organization, United Nations; American Philosophical Society; William Waterman Fund of the Research Corporation of America; National Library of Medicine; Pacific Science Board, National Research Council; U.S. National Museum, Smithsonian Institution. For their many and varied contributions, we are deeply grateful to all of these organizations and their staff.

A complete list of individuals who have contributed materially to this project would probably fill a book greater in size than the present tome. Their names have thus been regretfully omitted. To the listed and those unlisted, however, we are deeply grateful for their highly valued contributions. To the adventurer we wish pleasant diving; while those who prefer the security of their living room will, we believe, find this book both interesting and profitable reading.

Bruce W. Halstead, Paul S. Auerbach and
Dorman R. Campbell

NOTE: The author names that follow the scientific names of specimens are sometimes included in parentheses and on other occasions are not. The taxonomic significance of this is that where an author's name appears without parentheses it indicates that the scientific name was described by the author as presented. But if an author's name is included within parentheses, this indicates that a subsequent taxonomist has modified the name and placed the species in a different genus.

CHAPTER 1

TRAUMATOGENIC ANIMALS

Traumatogenic animals are those which could possibly cause a wound of some type. These marine animals could injure with a bite, sting, puncture, or merely because of their large size. They have been identified here because of the potential threat they present to humans. Included in this chapter are sharks, rays, barracudas, moray eels, needlefish, sawfish, groupers, saltwater crocodiles, seals and sea lions, as well as killer whales, polar bears and giant clams.

SHARKS

Phylum CHORDATA
Class CHONDRICHTHYES
Subclass ELASMOBRANCHII

Sharks are the most feared animals in the oceans of the world. With the great number of publications and films dealing with shark attacks, real and unconfirmed, the public awareness of sharks, and in many cases an unjustified fear of them, has reached an all-time high. During the past 40 years, a great deal of valuable information on the biology of sharks has come to light. However, scientists continue to suffer from a general lack of knowledge about these magnificent creatures. The following section is a summary of the important facts concerning sharks and their habits. Much of this data is based on investigations reported in the *International Shark Attack File* (US Navy Bureau of Medicine and Surgery, 1973). This file was developed by a group of distinguished scientists and is a collection of reports of shark attacks from all over the world.

There are about 350 species of sharks known to ichthyologists, but only 32 species have been definitely known to attack man. Dangerous sharks are found in most of the oceans of the world. However, most shark attacks have occurred between latitudes 47° south and 46° north. The most northerly incident happened in the upper Adriatic Sea, while the most southerly attack took place off South Island, New Zealand. Fifty-four per cent of attacks have taken place south of the equator; none has been reported in the Arctic or Antarctic regions. Shark attacks are most common in warmer and tropical waters, where the water temperatures are above 20°C (68°F). They usually occur during the warmer seasons, in late afternoon and at night.

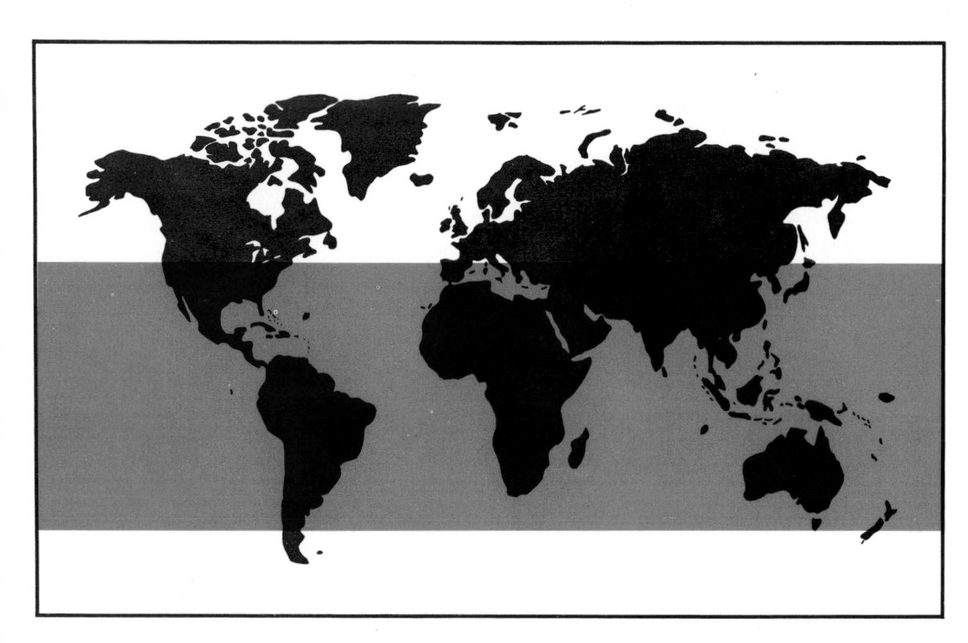

Shark attack zones

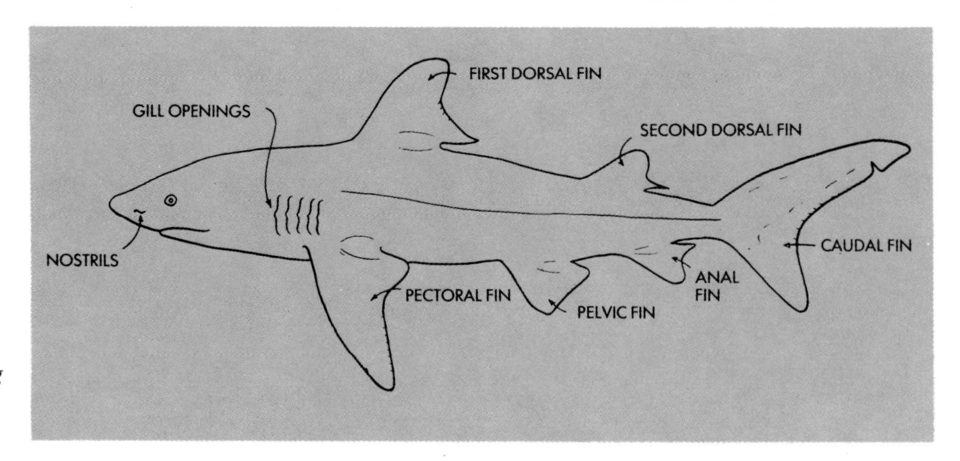

Generalized drawing showing typical external anatomy of a shark

The following paragraphs discuss the shark's body and senses, concentrating on its adaptions for hunting and feeding and the resultant dangers presented to humans.

Sharks vary greatly in size. Fully grown specimens can range in length from a mere 15 cm (6 inches) to more than 15 m (50 feet), depending on species. An example of a larger shark is the giant whale shark which can reach a length of 15 m. This species eats mostly plankton and its 3000 to 5000 teeth are very small and not suitable for biting or chewing larger objects, including humans.

A shark's jaws contain several rows of teeth. The number of rows of teeth in use at any one time varies from one to five, depending on the species of shark. If one looks at the inside of a shark's jaw, several rows of replacement teeth can be seen lying in a reversed position. These extra rows of teeth point upward in the upper jaw and downward in the lower jaw. Whenever a tooth is lost, a new one moves forward from the back to replace it. Hence, in the normal course of events, there is no chance that the shark will ever be without a full set of teeth. An instrument called a shark bite meter has been used to measure the actual force of a shark bite. On testing, some of the larger sharks have been found to bite with a force of 18 metric tons per square inch of tooth surface (about 20 tons per square inch).

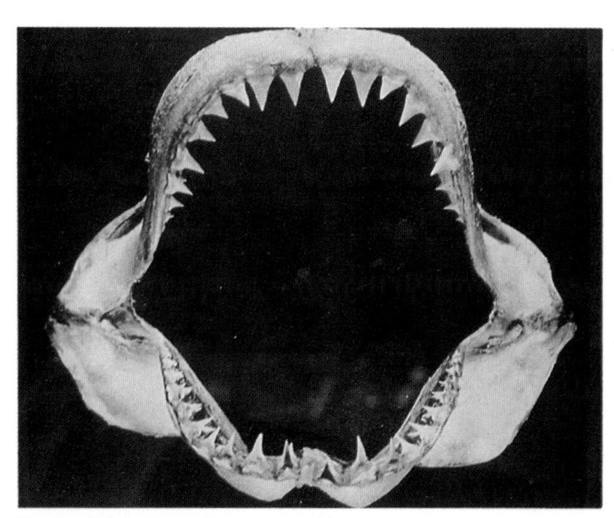

Jaws of ***Carcharodon carcharias*** *(great white shark)*

Scientists who once believed that sharks could not see very well, now know that sharks can see color and are able to see objects moving past a darkened background. Sharks depend on their eyes to perceive objects at a distance of 15 m (50 feet) or less, depending on the clarity of the water. Even though a shark may be twisting, turning or swiftly moving through the water, its strong eye musculature permits it to maintain a constant field of vision. Some sharks spend much of their time in very deep water where there is little light. Because of this, some have a special mirror-like layer, the tapetum lucidum, situated under the retina. This reflects incoming light back through the retina, permitting optimum utilization of light in a dimly lit environment.

Sharks have an impressive sense of smell. In fact, it has been found that a major portion of a shark's brain is used for olfaction. As previously mentioned, sharks rely on their eyes to perceive objects at distances less than 15 m (50 feet). For distances greater than this, sharks may rely on their olfaction, which allows them to detect food or blood in the water at levels of only a few parts per hundred million.

Sharks are able to taste things and tell the difference between food and other objects. Their taste buds undoubtedly play an important role in determining what they eat.

There are several sensory systems in the shark which can detect vibration in the water and may be called hearing mechanisms. These include the inner ear, the lateral line, and the ampullae of Lorenzini.

Frenzied feeding pattern

The ear position is directly behind the eye. Although sharks do not have an external ear, there is an inner ear which is capable of detecting vibrations from thousands of meters away. This, with olfaction, is one of the first mechanisms used by the shark in detecting food.

The lateral ear extends down each flank of the shark. The line has specialized tubes filled with mucus which can sense vibrations in the water. Even small ripples may be detected.

The ampullae of Lorenzini are sensory organs which are visible as small pores on the head of a shark. They are sensitive to temperature, salinity, electrical impulses and pressure (touch). These pore-like openings, located especially around the snout region, ooze transparent mucus when pressed. They were first described in detail by Stephano Lorenzini in 1678.

The hunting practices of sharks are largely controlled by a combination of these senses. Because of them, sharks are able to detect very slight movements in the water. They can sense or hear struggling fish or swimming people from great distances. This is why a swimmer must remember that sharks can detect them from far away and are more likely to be attracted to jerky movements (a stress reaction in fish) rather than smooth, slow movements.

Sharks have chemoreceptors located in the skin which make it possible for them to determine if there are harmful chemicals in the water, and to gauge salinity and other chemical changes in the water. Scientists are hopeful that a better understanding of this function could lead to the development of an effective chemical shark repellent.

The feeding patterns or habits of sharks vary substantially. In the normal feeding pattern the movements of the shark usually appear slow and determined. At other times they are jerky and fast. The swimming pattern, approach and final attack can vary with the type of shark and the particular situation.

The frenzied feeding pattern usually occurs in the presence of large quantities of food, or following a catastrophe, such as an explosion, the sinking of a vessel or the crash of a plane into the water. The frenzied feeding pattern is further encouraged if sharks congregate in large numbers. This pattern is very irregular and does not conform to the normal feeding movements. During a frenzy, sharks have been seen swimming straight up to the surface of a lagoon, snapping at floating objects and then suddenly plunging, banging their noses on the bottom while snapping savagely at anything in sight. It is during periods such as these that most shark repellent devices become ineffective and may even make matters worse. In one instance, cannabalism was observed in the gray reef shark (*Carcharhinus amblyrhynchos*). This frenzied behavior was stimulated by an underwater blast which killed a school of snappers. While one shark fed on a snapper that had been caught under some coral, a second shark moved in and with one quick bite ripped out the belly of the first shark. Within a few moments the helpless shark was completely eaten by other members of the school which had joined the killing.

Typical shark attack posture

Two views of agonistic display

***Carcharhinus amblyrhynchos* (Bleeker)**

ATTACK BEHAVIOR

In many instances attacking sharks are not seen by the victim before the attack. In other cases, aggressive behavior has been observed by the victim or onlookers. Sharks have been seen swimming back and forth before making a rush at the victim. The shark's pectoral fins have been observed pointing downward, with the shark's back in a humped position. Prior to attack, a shark might swim stiffly with its whole body, its head shaking in much the same fashion as its tail. This activity, which has been seen just before a swift pass at the victim, is called agonistic or combative behavior.

As the shark approaches the actual moment of attack, gentle bumps or violent collisions into the victim's body have been reported. These collisions have been known to knock the victim completely out of the water.

Shark attacks are very unpredictable, and so it is difficult to say exactly what the attack approach will be. For instance, sharks have been known to pick out one person in a crowd of swimmers and, attracted by blood of the damaged individual, continue to attack that person and no one else, including those trying to rescue the victim. At other times, sharks have been observed swimming disinterestedly near divers and actually coming close to look at the human intruders, and then swimming away. This unpredictable element of shark behavior, together with many species' awesome size and potentially fierce attitude, has helped the creature to gain its feared reputation.

A significant number of shark attacks have happened because of provoked encounters, such as spearing, poking at the shark, grabbing it by the tail, offering fish to the shark, blocking off an escape passage, or through some other annoyance. Needless to say, actions of this type should be avoided. Sharks are powerful and can move extremely quickly. Even a relatively small shark can inflict serious injuries when provoked, and a group of small sharks can do more damage than a single large one.

It is important to realize that unprovoked attacks have taken place when the victims appeared to be minding their own business. It is therefore essential to remember that there are certain things which are known to encourage sharks to attack (see Shark Attacks, Prevention, page 25).

The following sharks have been implicated in attacks on humans. The list of families is arranged in alphabetical order. Drawings of teeth are included as an aid in species identification.

Family CARCHARHINIDAE

This is the largest family of sharks, with at least 13 genera and many species. Many members are dangerous and several have caused human deaths.

Carcharhinus albimarginatus **(Rüppell)**
Silvertip shark

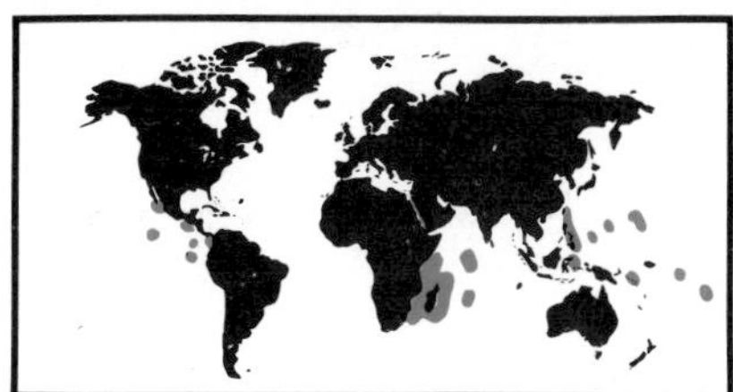

DESCRIPTION Length up to 2.7 m (9 feet); dark gray with strikingly conspicuous white tips and posterior margins on all fins.

HABITAT Abundant around reefs and off-shore islands, from the surface to depths of between 600 and 800 m (180 and 240 feet). Widespread throughout the warm waters of the Pacific and Indian Oceans.

OTHER POINTS This species is thought to be very aggressive toward one another, and individuals often have evidence of battle scars.

Carcharhinus amblyrhynchos **(Bleeker)**
Gray reef shark

DESCRIPTION Length 2.1 m (7 feet) or more; grayish above, paler to whitish below. Fins are gray to black but not black tipped.

HABITAT Abundant in lagoons and around coral reefs in the Indo-Pacific and Indian Oceans.

OTHER POINTS Reported in attacks on humans. Frequently found in large schools. May become very aggressive, entering into a feeding frenzy when food is in the water.

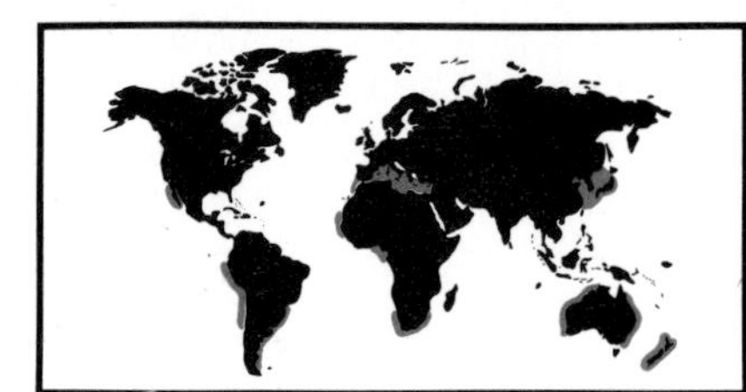

Carcharhinus brachyurus **(Günther)**
Bronze whaler, copper shark

DESCRIPTION Length up to 2.7 m (9 feet); bronze to golden above, white below.

HABITAT Open sea and along the coasts of New Zealand.

Carcharhinus galapagensis **(Snodgrass and Heller)**
Galapagos shark

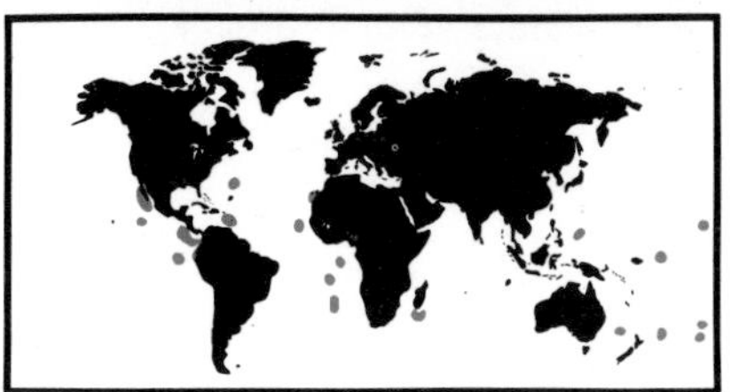

DESCRIPTION Length up to 2.4 m (8 feet); gray above, cream below; often mottled with gray, with tips of pectoral and first dorsal fins darker.

HABITAT Shallow water species found in the West Indies and in the eastern Pacific westward to the Indian Ocean.

OTHER POINTS One of the most abundant shallow water species of shark in the eastern Pacific.

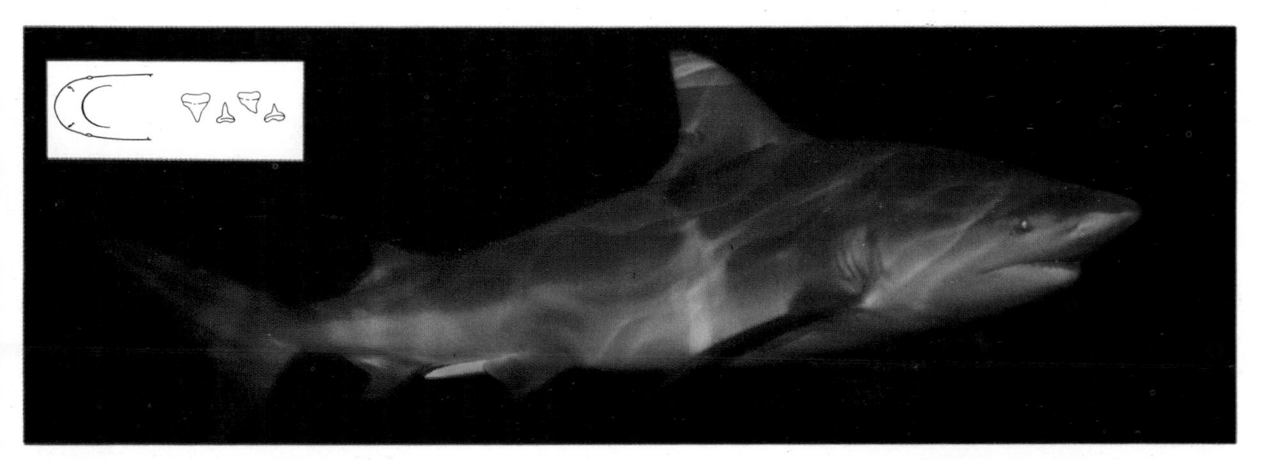

Carcharhinus leucas **(Valenciennes)**
Bull shark

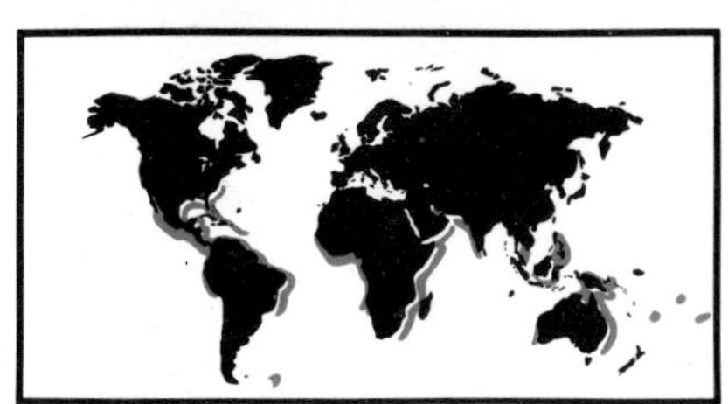

DESCRIPTION Length up to 3.6 m (12 feet); gray above, white below. Fins of younger specimens are dark tipped.

HABITAT Although mainly located in the warm oceans of the world, has been found in Lake Nicaragua, Lake Izabel (Guatemala), Lake Jamoer (New Guinea) and the Amazon River as well as the rivers of Guatemala, Australia, Iraq and southeast Africa as far as 1600 km (1000 miles) inland.

OTHER POINTS Known locally by many different common names (Lake Nicaragua shark, Zambezi shark, Ganges River shark). Has repeatedly been implicated in attacks on humans. Slow swimmer, except when stimulated by food.

Carcharhinus longimanus (Poey)
Whitetip oceanic shark

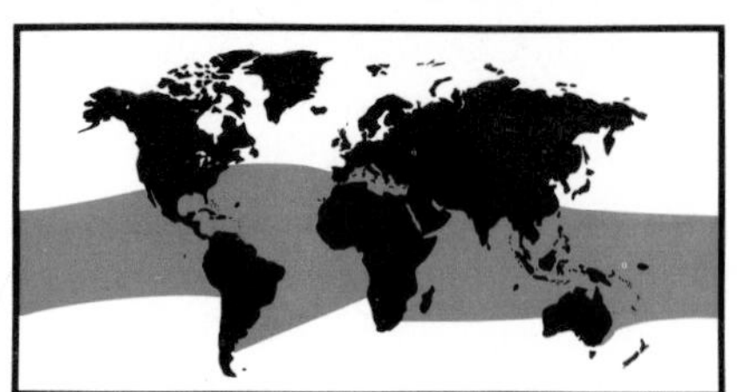

DESCRIPTION Length up to 3.6 m (12 feet); blue, gray or brown above, yellow or white below. Fins are white tipped (sometimes dusky-tipped in younger specimens) mottled with gray. First dorsal fin with wide, rounded top; lower part of caudal fin convex shaped. Short snout in front of nostrils.

HABITAT Oceanic species that seldom if ever swims close to shore. It is widely distributed throughout the warm waters of the Atlantic, Pacific and Indian Oceans.

OTHER POINTS Very active, seemingly fearless.

Carcharhinus melanopterus (Quoy and Gaimard)
Blacktip reef shark

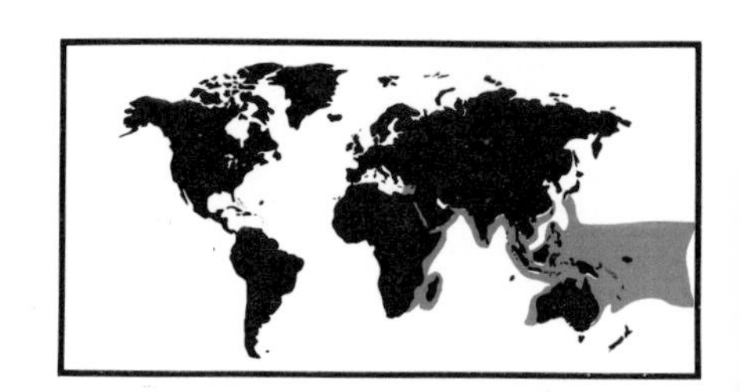

DESCRIPTION Length up to 1.8 m (6 feet); brown to black above, paler to white below. Fins usually black.

HABITAT Very abundant in the western Indo-Pacific, living in shallow water of coral reefs. Common in shallow water lagoons and estuaries. Lives in confined areas.

OTHER POINTS Although this species has been known to attack humans, it can usually be frightened away by making overt movements towards it.

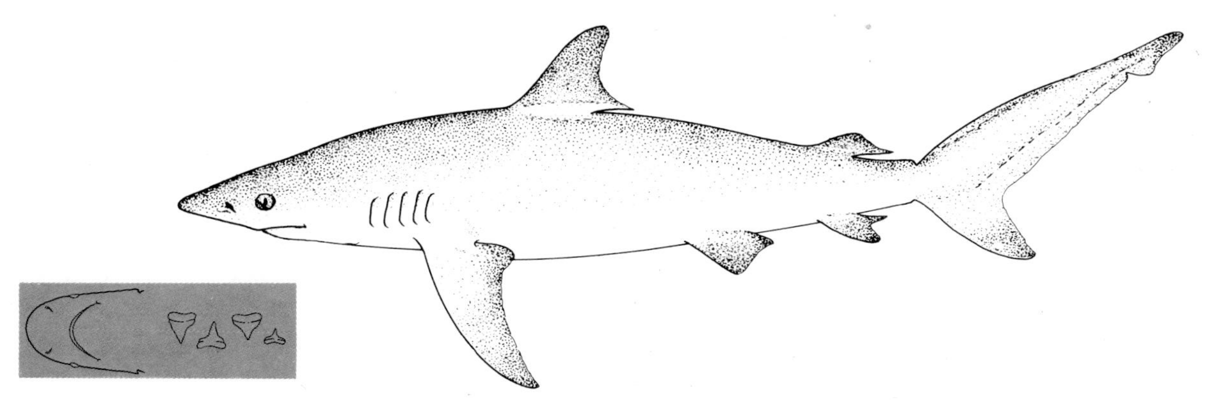

***Carcharhinus obscurus* (LeSueur)**
Dusky shark

DESCRIPTION Length up to 3.6 m (12 feet); blue-gray to pale gray above, white below. Undersides and tips of pectoral fins are grayish to dusky.

HABITAT Lives mainly in the open ocean, but may enter shallow coastal areas. Has been recorded from the warm Atlantic, eastern Pacific and western Indian Oceans.

OTHER POINTS Migrates seasonally, moving north during warmer summer months and retreating south when the waters cool.

***Galeocerdo cuvier* (Peron and LeSueur)**
Tiger shark

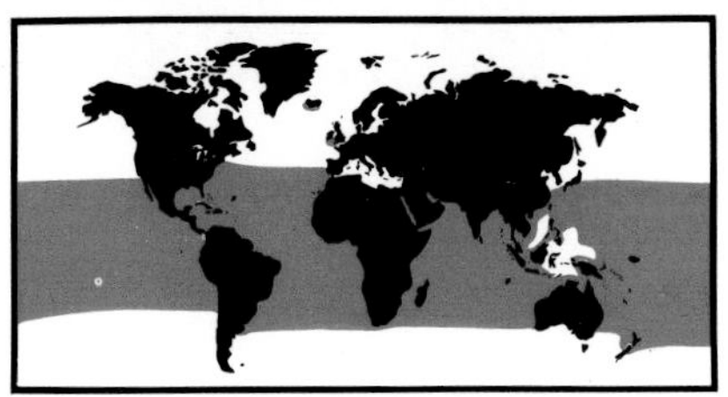

DESCRIPTION Length up to 5.4 m (18 feet), but most specimens range from 3.6 to 4.5 m (12 to 15 feet). Curved notched teeth, short snout and sharply pointed tail. Gray or grayish brown, darker above than on the sides and underside. Back and fins of younger specimens are striped and blotched; older and larger specimens are patterned on the tail only, or not at all.

HABITAT Near the shore and offshore; may also enter river mouths. Widespread throughout tropical and subtropical regions of all oceans.

OTHER POINTS One of the most common of the large sharks in the tropics, it is second only to *Carcharodon carcharias* (great white shark) in recorded attacks on humans. At least 27 documented attacks are sourced to it. Generally a slow swimmer, becomes vigorous when provoked or when stimulated by food or blood in the water.

***Negaprion brevirostris* (Poey)**
Lemon shark

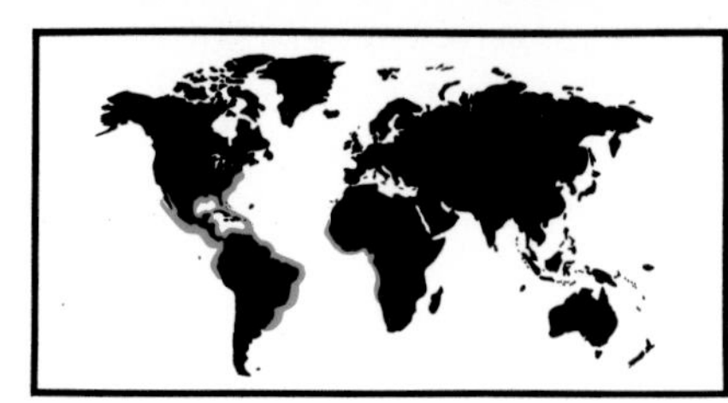

DESCRIPTION Length up to 3.3 m (11 feet); yellowish-brown to dark brown above, yellowish sides, white to yellowish or olive gray below. Unique teeth and broadly rounded snout. Second dorsal fin almost as large as the first.

HABITAT Primarily an inshore species, commonly found in saltwater creeks, bays and sounds and around docks. Ranges from New Jersey to northern Brazil; also found in tropical west Africa but is often most commonly located in the waters of Florida and the Keys.

OTHER POINTS Has been studied extensively as a consequence of its ability to live well in captivity.

***Prionace glauca* (Linnaeus)**
Great blue shark

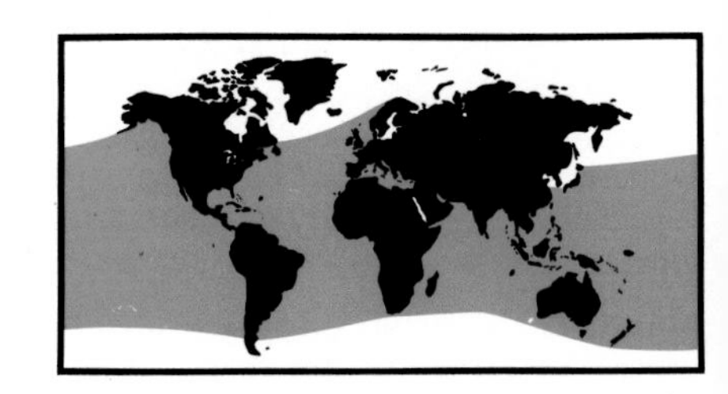

DESCRIPTION Length up to 3.6 m (12 feet) or more; indigo blue above, slightly lighter on sides and white below.

HABITAT Lives mainly in the open sea but may be found along coastal areas throughout tropical and warm temperate oceans.

OTHER POINTS Slow swimmer but becomes active and aggressive when stimulated by food or blood in the water. The photograph here shows the results of an experiment when dead fish were packed inside a chain mail suit in order to entice an attack.

***Triaenodon obesus* (Rüppell)**
Whitetip reef shark

DESCRIPTION Length up to 2.4 m (8 feet); gray above, white below. White markings on the first dorsal, pectoral and upper caudal fins.

HABITAT Around reefs, coral atolls and along shorelines throughout the warm waters of the Pacific and Indian Oceans, and the Red Sea.

OTHER POINTS Sometimes confused with the whitetip oceanic shark, although they differ in form and color and the latter is a deep-water species. Rarely reported to attack humans, but is considered to be potentially dangerous.

Family GINGLYMOSTOMATIDAE

This family consists of about 11 genera with between 12 and 28 species recognized. They are characterized by the presence of fleshy nasal barbels just anterior to the mouth. They are slow moving, bottom-dwelling sharks but nevertheless must be regarded as dangerous.

***Ginglymostoma cirratum* (Bonnaterre)**
Nurse shark

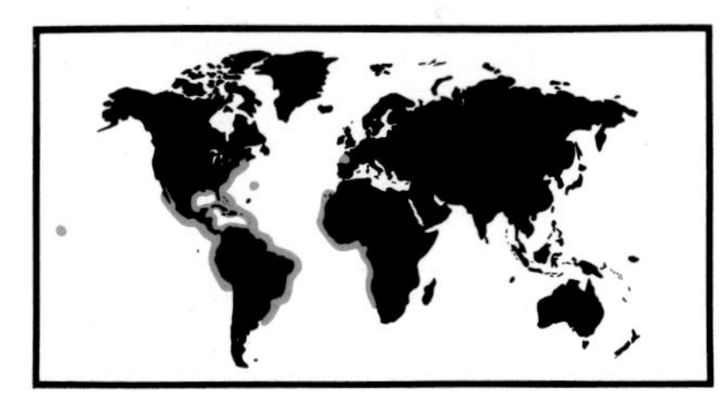

DESCRIPTION Length 4.2 m (14 feet); yellowish-brown to grayish-brown with or without small dark spots.

HABITAT Common inshore bottom-dwelling shark with nocturnal habits. Often found lying on sea bed in schools. Found throughout the warm waters of the Atlantic and Pacific Oceans.

OTHER POINTS May attack humans if they are molested, accidentally stepped upon or too closely approached. Closely related species are found in the Indian Ocean.

Family ISURIDAE

This family consists of three genera: *Carcharodon*, *Isurus* and *Lamna*. Within these genera there are around half a dozen species. All members of this family are large and the following three species are potentially the most dangerous.

***Carcharodon carcharias* (Linnaeus)**
Great white shark

DESCRIPTION Length up to 6 m (20 feet) or more; grayish-brown, gray or blueish above, off-white below. There is usually a black spot on the body at the rear of the pectoral fin base. Outer tips of the pectoral fins are black. Blunt, cone-shaped head, heavy body, strongly crescent-shaped dorsal fin, large triangular notched jagged teeth.

HABITAT Generally open ocean, but does swim into shallow water. Most attacks occur from estuaries. It ranges into the tropical and temperate oceans of the world.

OTHER POINTS Usually swims alone. Reported by some experts to attack humans as a result of mistaking them for its normal prey of seals.

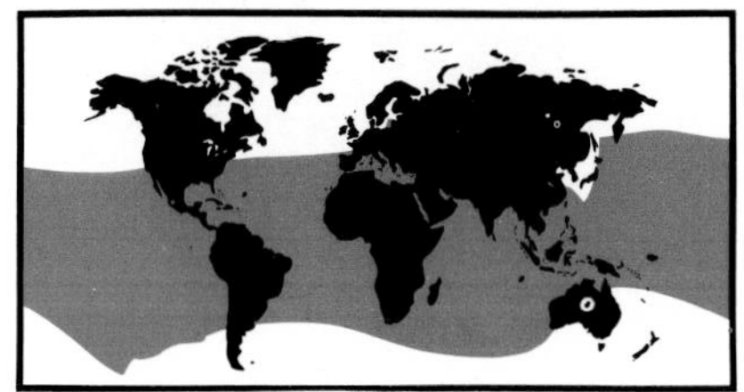

Isurus oxyrinchus **(Rafinesque)**
Mako shark, sharp nosed mackerel shark, blue pointer, bonito

DESCRIPTION Length 3.6 m (12 feet); dark blueish-gray or blue above, white below.

HABITAT Mainly open ocean, but may come close to shore. Inhabits all tropical and temperate oceans.

OTHER POINTS An aggressive and dangerous shark with fearsome-looking jaws. One of the most active species; when hooked on a fishing line may leap out of the water, and has been reported to attack boats.

Lamna nasus **(Bonnaterre)**
Mackerel shark, porbeagle, salmon shark

DESCRIPTION Length 3 m (10 feet) or more; bluish-gray or gray above, white below; first dorsal fin with a conspicuous white rear tip.

HABITAT A littoral and epipelagic shark preferring water temperatures less than 16°C (60°F). Tends to occupy temperate regions of all oceans.

OTHER POINTS Slow swimmer, but active and aggressive when in pursuit of food.

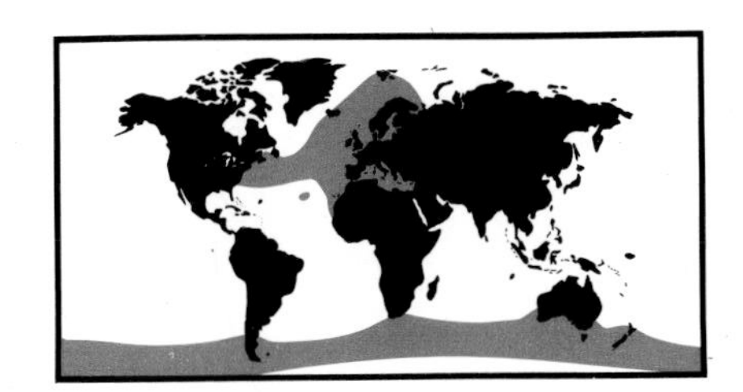

Family ODONTASPIDIDAE

This is a small family with two genera, *Odontaspis* and *Pseudocarcharius*, and five to eight species. These are large sharks of warm temperate waters throughout the world. In some areas they have a fearsome reputation as being extremely dangerous to man, but this is primarily a result of confusion with other species.

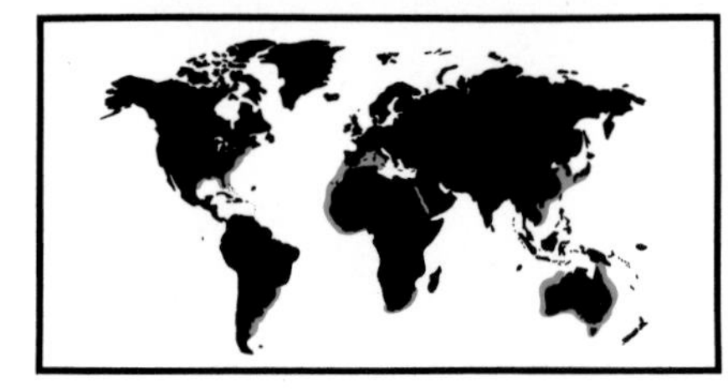

Odontaspis taurus **(Rafinesque)**
Sand shark, gray nurse shark

DESCRIPTION Length up to 3 m (10 feet); gray or brown above, paler to dirty white below, sparsely flecked on the back and sides. Fins with brownish spots.

HABITAT Lives mostly near the bottom as an inshore species throughout the Mediterranean, the Atlantic, and the Indian Ocean along eastern Africa.

OTHER POINTS Although comparatively sluggish, it has a voracious appetite, feeding very actively at night.

Family ORECTOLOBIDAE

This is a large family with about 12 genera and numerous species found in the Indo-Pacific and Red Sea. All species are relatively small in size and live on reefs and in shallow water. Several species have been reported to be involved in attacks on humans, but if unmolested they are harmless.

Orectolobus maculatus **(Bonnaterre)**
Wobbegong

DESCRIPTION Length up to 2.1 m (7 feet); grayish and brownish blotches, spots and marbling together with fleshy lobes for camouflage.

HABITAT Shallow water, especially weed covered rocks in eastern and southeastern Australia.

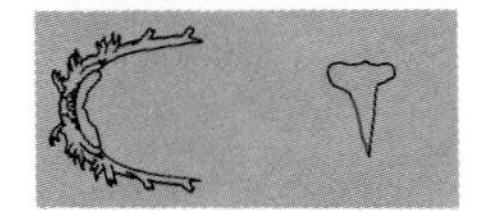

Orectolobus ornatus **(Devis)**
Tasselled wobbegong

DESCRIPTION Length up to 3.6 m (12 feet); reticular pattern of narrow darker lines on a light background with scattered large dots above. Abundant branched fleshy lobes below mouth opening.

HABITAT Inshore bottom species often found on coral reefs.

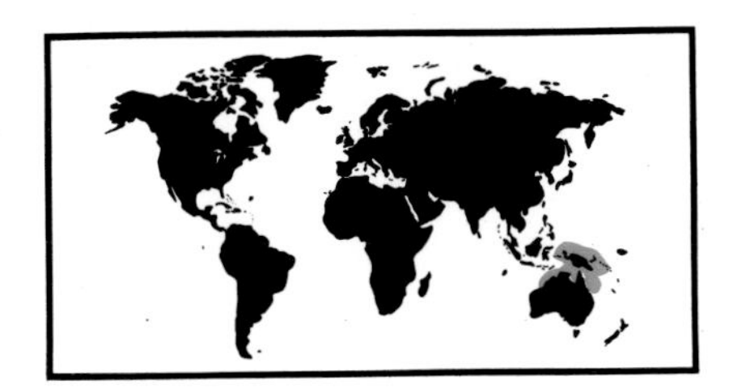

Family SPHYRNIDAE

The hammerhead family of sharks is best known for its peculiar flat, wide, hammer or bonnet-shaped head, which sets them apart from all other sharks. They are found in warm waters of all oceans. A number of attacks on humans have been reported, but the exact species involved are unknown.

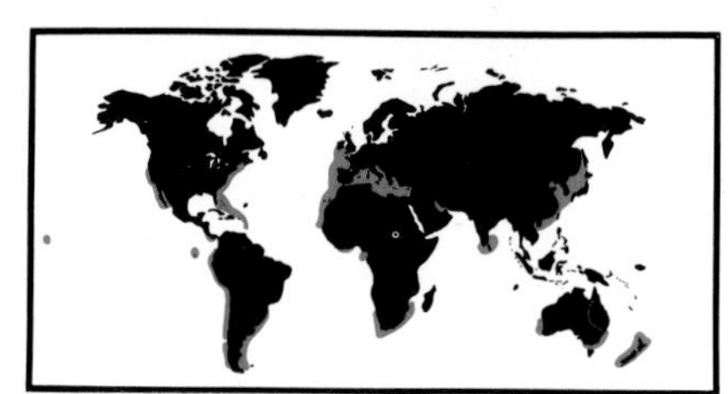

Sphyrna zygaena **(Linnaeus)**
Smooth hammerhead shark

DESCRIPTION Length 4.2 m (14 feet); dark olive or brownish gray above, paler to white below; tips of fins dusky to black.

HABITAT Frequently found swimming at the surface of the open ocean. A common species found in tropical and warm temperate waters of all oceans.

OTHER POINTS Easily recognized by head shape; a leisurely swimmer in general.

SHARK ATTACKS

Prevention

Most shark attacks (about 86%) are believed to be unprovoked. Nevertheless, a person should make every effort not to antagonize sharks. They are attracted by brightly colored or shiny metallic objects, blood in the water, food (dead fish), low frequency vibrations and explosions. They are also attracted by spear fishing; a dying fish struggling on the end of a spear is a powerful invitation to sharks.

Avoid murky water inhabited by sharks in which there is poor visibility. Garbage is known to attract them; therefore, avoid swimming in estuaries where garbage is dumped. Use caution when swimming during late afternoon and at night in areas where sharks are most apt to be feeding.

If you encounter a shark, move with slow purposeful movements. If a large shark appears to be too inquisitive, make every effort to get out of the water, but *do not panic*. Keep your eyes on the animal and never turn your back. Remember that you are most vulnerable at the surface. If you are in SCUBA gear and have adequate air supply, submerge, unless exit from the water is easily managed. Hide behind a coral head, in a crevice, or with some solid object at your back. Face your adversary and remain as quiet as possible. Use an available object to fend off the intruder. A hard thump on the nose or a poke in the eye is often helpful in discouraging a shark. Never dive alone.

Sharks are attracted to even small amounts of blood. Therefore, when spear fishing, never tie injured, bloody fish to your body. Toss the fish into a boat or raft, or anchor them some distance from where you are swimming. All sharks have skin covered with denticles that can severely abrade the human skin, so avoid brushing up against sharks; any resulting blood can trigger frenzied behavior and draw other sharks to the area. If you are wounded or bleeding for any reason, do not stay in the water. For this reason, women should probably not enter shark-infested waters during their menstrual period.

Treatment

Although the shark bite wound is usually from a single bite, it is often extensive, with jagged edges, multiple crescentric linear cuts, and components of laceration, avulsion and compression, depending on the type of shark. Bleeding may be extensive, particularly if large blood vessels or highly vascular internal structures are damaged. Because of the force involved, injuries may extend internally into the abdomen and thorax where wounds occur on the body.

First aid begins with rescue of the victim from the water. The success or failure in administering first aid depends primarily upon the extent of the injury and the rapidity with which therapy can begin in the treatment for shock. Most shark fatalities are the result of massive tissue loss and hemorrhage or of panic and drowning. In rare instances it may be necessary to partially control the bleeding while the victim is still in the water by the application of manual compression directly over the wound. The victim should be removed from the water as soon as possible to avoid further attack, control bleeding, prevent shock and diminish the effects of hypothermia. Any shark bite, whether it appears to be minor or severe, must be taken seriously.

Initial treatment should begin on the beach or in the rescue boat. Place the victim in the head-down position. Anticipate a serious injury and expedite transport to an appropriate trauma center as soon as possible. Do not attempt exploration of the wound yourself.

Although the possibility of further contamination of the wound during this initial period is small, some precautions will minimize the infection risk. The wound is already contaminated with ocean water, sand and possibly the shark's teeth, and therefore also with marine microflora. The major pathogens include virulent genera, such as *Vibrio, Erysipelothrix, Pseudomonas* and *Clostridium,* as well as scores of other organisms. If possible, the wound should be promptly irrigated with sterile saline, lactated Ringer's or water, using non-sterile fresh (non-marine) water if surgical irrigation fluid is not available. If transport to a hospital is likely to take more than 12 hours, the victim should be given oral ciprofloxacin (750 mg) or else trimethoprim-sulfamethoxazole (160/800 mg) every 12 hours.

It is essential to control the bleeding in order to avoid hypovolemic shock. All means available should be used to apply compression dressings. For a pressure bandage, use whatever is available: gauze cotton, clothing, towels, sack cloth, etc. Even if a large blood vessel has been severed, bleeding can usually be controlled with direct pressure, which allows vasospasm and early clot formation. If traumatic amputation has occurred, bleeding may be profuse. If bleeding from a visible major artery or vein threatens the victim with exsanguination, the vessel may be directly ligated, or a proximal pressure point may be compressed. Never attempt to blindly clamp vessels, as this is rarely effective and frequently leads to inadvertent and irreparable nerve damage. The decision to ligate a large blood vessel or to apply a tourniquet is essentially one to sacrifice a limb in order to save a life. If a constricting band is used to occlude arterial blood flow, apply direct pressure to the wound simultaneously. Every 10 minutes, loosen the tourniquet briefly to see if direct pressure alone will control the bleeding.

If possible, intravenous fluids should be administered to the patient en route to the hospital, particularly if transport is likely to be prolonged or delayed. If volume must be replaced in large quantities, at least two large-bore intravenous lines should be inserted in the uninvolved extremities. Administer lactated Ringer's, normal saline, plasmalyte, hetastarch, hypertonic saline-dextran or group O Rh-negative blood until type-specific or fully cross-matched blood products are available. Administer oxygen by face mask at 5 to 10 liters per minute. If necessary a three-compartment MAST suit should be applied, provided there will be no excessive hemorrhage from areas not covered by the garment.

Keep the patient warm, but do not smother him in heavy blankets on a hot beach under a boiling sun. Do not give the victim liquids by mouth if he has altered mental status or if it is likely that he will soon go to the operating room. If the patient is cardiovascularly stable, naloxone-reversible narcotics may be administered in increments for pain relief. Reassure the victim frequently. For further medical advice see Chapter 7.

SHARK REPELLENTS

Name	***Description***	***Comments***
US NAVY SHARK REPELLENT *Shark Chaser*	A black dye and lead acetate powder which is spread in the water near divers.	Does not repel all types of sharks.
SHARK BILLY	A stick tipped with a sharp metal object, such as a nail.	Has been found useful to push a shark away.
BANG STICK *Smokey or Powerhead*	An explosive device which delivers a shot or slug to the brain or other vital organ of the shark. Firing is done by forcefully driving the triggering mechanism into the shark.	This is a very dangerous weapon, which is only effective when dealing with one shark. Its use is highly controversial.
CO_2 GAS DART GUN	A CO_2 charge is fired into the abdomen of the shark causing the animal to lose control of its movements. It then floats to the surface.	These are not always effective because of the difficulty in penetrating a shark's hide.
SONIC VIBRATION DEVICE	Instrument which produces sounds that repel sharks.	These are being tested at the time of writing.
SHARK SCREEN	A water filled plastic sack suspended from the surface by flotation rings. The sack is large enough to contain a floating adult.	It provides a limited amount of protection by means of camouflage, and prevents blood from a wound from entering the water.
MESHING	Large gill nets are submerged along the beach close to the surf line. Sharks are caught in the net, where they die.	This has been the most effective method to date for controlling beach areas inhabited by dangerous sharks.
MISCELLANEOUS EXPERIMENTAL DEVICES	a Electrical and sonic devices. b Chemical agents. Toxins–pardaxin from the Moses sole, *Pardachirus marmoratus*; and surfactants such as sodium dodecylsulfate, a potent detergent and foaming agent.	Experimental but promising.

RAYS

Class OSTEICHTHYES
Order RAJIFORMES
Family MOBULIDAE

There are two genera in this family, comprising about 10 species.

***Manta birostris* (Dondorff)**
Giant devil ray or manta

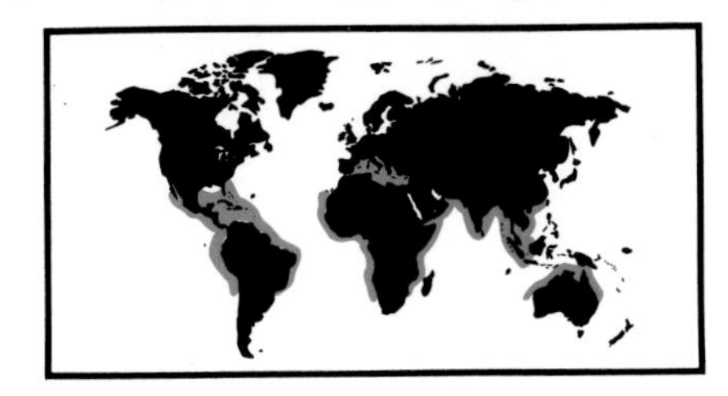

DESCRIPTION Spread of 6 m (20 feet) or more, weight 1500 kg (3300 pounds) or more; reddish or olive-brown to black above, light on the underside.

HABITAT Open ocean; generally seen swimming or basking near the water surface with the tips of their long pectoral fins curling above the surface. Found in tropical/subtropical belts in both hemispheres.

OTHER POINTS Mainly plankton feeders, mantas are not aggressive but are dangerous because of their rough skin and huge size which has led to the deaths of deep-sea divers whose air lines have been entangled with these creatures. Mantas can jump clear out of the ocean and have been known to tow small boats for many miles when hooked on a fishing line.

Family PRISTIDAE

This family has one genus, *Pristis*, with six species. The family is famous for its snout, which is produced in a long flat blade with teeth of equal size embedded in deep sockets on each side.

Pristis perotteti **(Müller and Henle)**
Sawfish

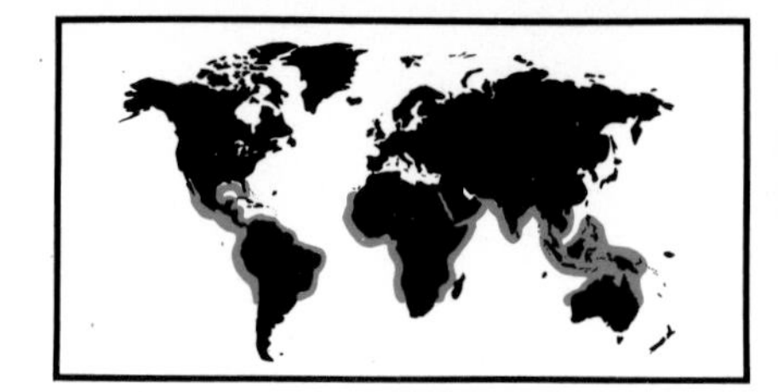

DESCRIPTION Length 5 m (16 feet), more or less uniformly brown, olive gray or yellowish above and on the sides, white below. A very large ray of shark-like appearance with an extremely pronounced snout forming a narrow and firm blade armed with transverse tooth-like structures.

HABITAT Shallow coastal waters, estuaries, river mouths and freshwater rivers and lakes in all tropical/subtropical areas, preferring a sandy or muddy bottom.

OTHER POINTS Has been found in the higher reaches of the Zambezi River.

BARRACUDAS

Order PERCIFORMES
Family SPHYRAENIDAE

There is one genus, *Sphyraena*, comprising 18 species, the great barracuda being the largest and best known.

Sphyraena barracuda **(Walbaum)**
Great barracuda

DESCRIPTION Length 1.8 to 2.4 m (6 to 8 feet), weight averages 48 kg (106 pounds); large mouth with relatively enormous teeth.

HABITAT Generally open tropical ocean. Widely distributed throughout tropical and subtropical waters of the world but most commonly found in the West Indies and Brazil north to Florida, and in the Indo-Pacific from the Red Sea to the Hawaiian Islands.

OTHER POINTS Rarely attack humans. Are attracted to anything entering the water, particularly brightly colored and silvery objects. Relying almost entirely on sight, may follow divers for hours. If they do attack they usually make one quick, fierce strike which, although serious, is rarely fatal.

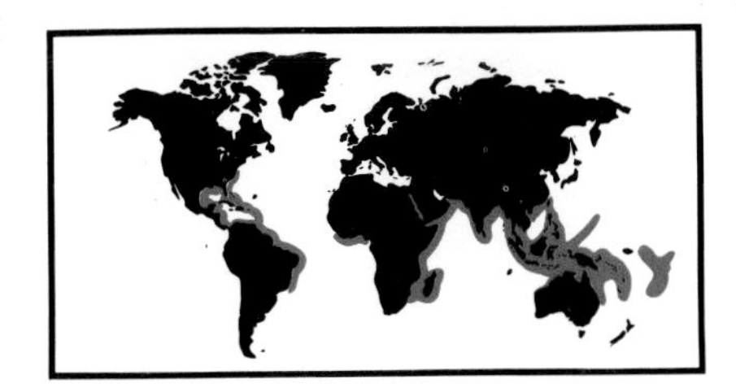

BARRACUDA ATTACKS

Prevention

Barracuda bites in man are quite rare, considering the frequency of encounters. Nevertheless, the barracuda is a voracious predator, highly feared in some parts of the world. Barracuda are attracted by bright shiny objects. Do not carry or tie dead fish around your body.

Treatment

Barracuda wounds tend to consist of straight V-shaped lacerations. Bleeding may be profuse. Treatment should be for deep lacerations, tissue loss, blood loss and shock, as for sharks (discussed on page 25).

MORAY EELS

Order ANGUILLIFORMES
Family MURAENIDAE

Moray eels have numerous sharp, fang-like teeth encased in a narrow, muscular jaw and are capable of inflicting a painful and crushing bite. The flesh of some species can also be poisonous. Moray eels may attack persons wading in shallow water over coral reefs, but seldom do so unless provoked. They are mostly active at night. When attacking, they have been known to hold on to their victim with a bulldog-like grip. The powerful and muscular bodies of these creatures are covered by a tough, leathery skin which is not readily penetrated by a knife. The slipperiness of the skin also makes if difficult to grasp the eel in case of attack.

Gymnothorax mordax **(Ayers)**
California moray eel

DESCRIPTION Length averages 1.5 m (5 feet); color variable, grayish, brownish, purplish or blackish, but most often with distinctive patterns of marbling, blotches or bands.

HABITAT Hide in crevices and holes, in coral or under rocks in reef flats along the California coastline.

OTHER POINTS A common California species, usually harmless unless aggravated.

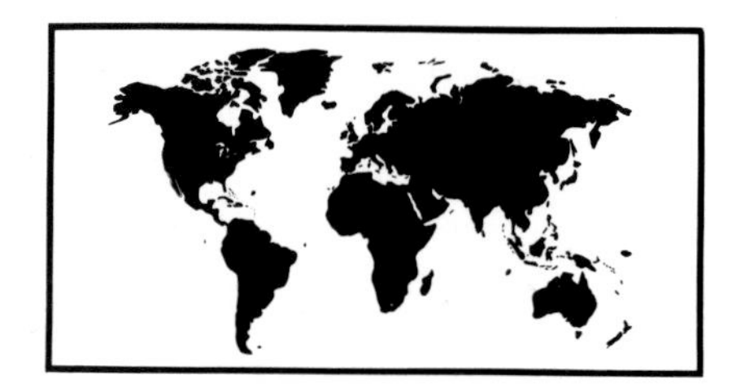

Muraena helena **(Linnaeus)**
Mediterranean moray eel

DESCRIPTION Length 1 m (39 inches); brown with large pale spots containing smaller brown spots forming a rosette pattern distinctive to this species.

HABITAT Shallow water rocky areas of the Mediterranean.

OTHER POINTS May have been used as a food since the days of the Roman Empire when they were specifically bred for human consumption.

MORAY EEL ATTACKS

Prevention

Moray eels rarely attack without provocation. Most eel bites are the result of divers placing their hands in holes and crevices inhabited by eels. Keep hands out of holes and wear gloves when contact is likely.

Treatment

Eels are not equipped with venomous fangs as are sea snakes. However, the palatine mucosae of some moray eels have been found to contain a toxin which may cause mild envenomation. The nature of the poison is unknown.

Eel bites tend to be either a series of puncture wounds made by the teeth or, rarely, ragged avulsion bite wounds. Bleeding may be profuse. First aid and treatment is similar to that for shark bites (see page 25). Infection from eel bites is often encountered.

NEEDLEFISH

Order CYPRINODONTIFORMES
Family BELONIDAE

There are about 10 genera included in this family, with 32 species. Some species are capable of making high jumps out of the water.

Tylosurus crocodilus **(Peron and Le Sueur)**
Hound needlefish, saltwater garfish

DESCRIPTION Length averages 1.8 m (6 feet); slender body with two long pointed jaws and sharp, unequal teeth.

HABITAT Bays, inshore areas and at times in deeper water; largely surface swimmers.

OTHER POINTS These animals are attracted by light at night. They can leap out of the water in the direction of the light and can also "leap-frog" over objects in the water. It is during these periods of excitement that they can cause puncture wounds with their sharp snouts.

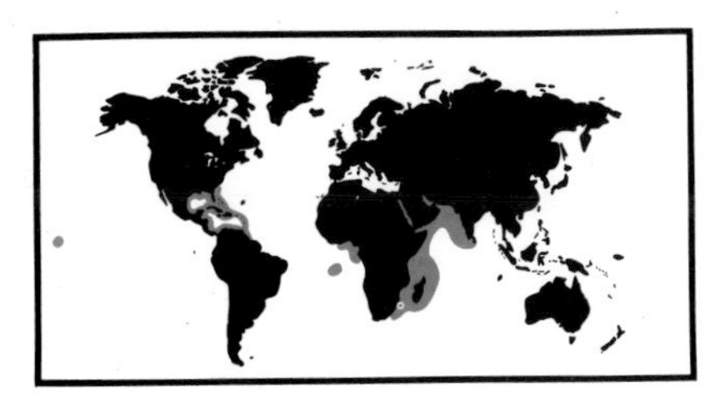

NEEDLEFISH TRAUMA

Prevention
Most needlefish injuries that have been documented have occurred to victims who have been fishing at night from small boats. The fish is attracted to the light and leaps out of the water, impaling any victim in its flight path.

Treatment
Needlefish injuries are of the puncture wound variety and require prompt surgical attention. Needlefish have been known to puncture the brain and chest and have caused fatalities. If a chest wound is present, cover the laceration with petrolatum gauze and apply pressure. Intravenous fluids may be required for hypotension. Administer oxygen by face mask or nasal cannula. Transfer the victim by ambulance to a trauma center as soon as possible.

GROUPERS

Order PERCIFORMES
Family SERRANIDAE

Groupers are considered here as a potential hazard mainly because of their possible large size, enormous jaws and fearless attitude. However, they are usually small, territorial reef-dwellers.

***Promicrops lanceolatus* (Bloch)**
Giant grouper, seabass, jewfish

DESCRIPTION Length 3 m (10 feet) or more; young specimens are yellowish or orange with four dark brown cross bands; adults are uniformly brown.
HABITAT Coral reefs, rock, or sandy areas in shallow waters of the Indo-Pacific.

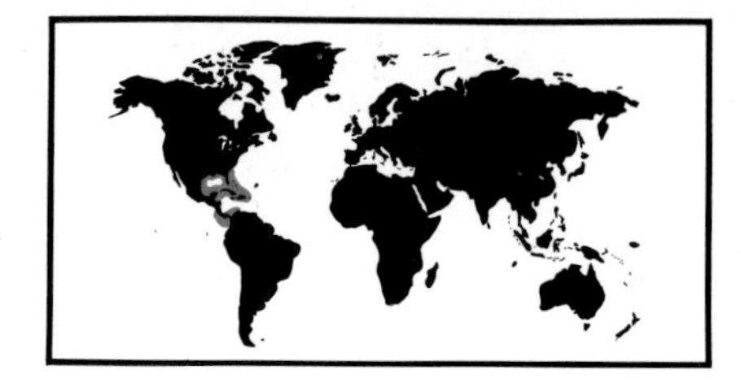

GIANT GROUPER ATTACKS

Prevention

Although they are generally not aggressive, large, angered specimens are capable of inflicting serious bites. Groupers tend to be quite territorial and may become aggressive when protecting their territory. Therefore it is always wise to examine an underwater cave before entering. Do not block the exit if a grouper is attempting to escape. Do not carry speared fish. Most scare tactics used against sharks are of no avail with groupers.

Treatment

The wounds may be ragged with extensive maceration of tissue. First aid and treatment procedures are similar to those employed for shark bites (see page 25).

SALTWATER CROCODILES

Class REPTILIA
Order CROCODILES (Crocodylia)
Suborder EUSUCHIA
Family CROCODILIDAE

The majority of all crocodilians belong to the genus *Crocodylus*, with 11 species. The saltwater crocodile, *Crocodylus porosus*, is one of the most dangerous of all marine animals.

Crocodylus porosus **(Schneider)**
Saltwater or estuarine crocodile

DESCRIPTION Length up to 6 m (20 feet) or more; brassy yellow, spotted or blotched with black above, whitish or pale yellow below; older specimens may be uniformly black above.

HABITAT Usually in coastal mangrove swamps, river mouths and brackish water inlets, but adult crocodiles have been seen swimming completely out of sight of any land. The large range of this crocodile is the result of its far-reaching trips out to sea. It ranges over an extensive geographical area including India, Sri Lanka, southern China, Malay Archipelago, Palau, Solomon Islands and northern Australia.

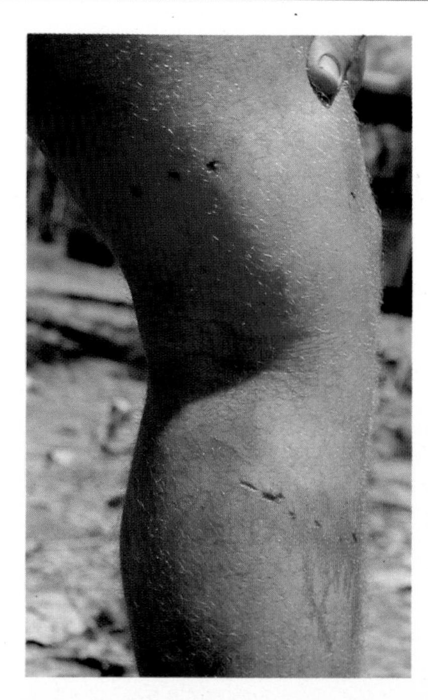

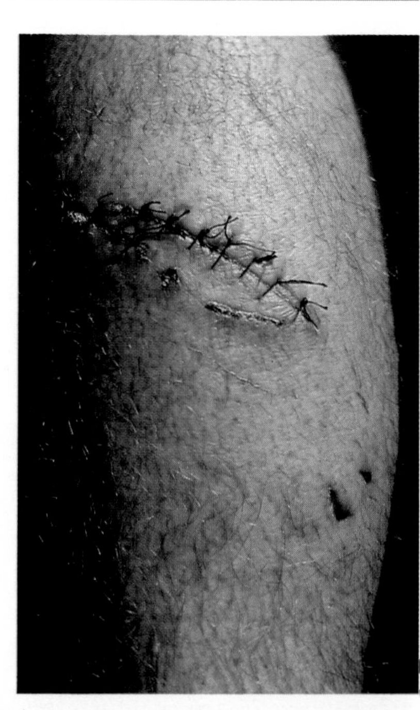

Right and far right: Bites from ***Crocodylus***
Below: ***Crocodylus porosus****, saltwater crocodile*

CROCODILE ATTACKS

Prevention

Crocodiles are extremely dangerous. They are ferocious and have taken numerous human lives in areas where they are endemic. They attack with a loud hissing noise, grasp the victim with their powerful jaws, and drown it with a quick twirling motion of the body. A crocodile may sweep a person off his feet with a swift blow of the tail.

Avoid swimming in murky brackish water inlets, river mouths, and mangrove swamps inhabited by saltwater crocodiles.

Treatment

Crocodile bites may result in massive tissue loss and profuse hemorrhage. Treatment may be complicated because of the concomitant submersion injury. Crocodile bite wounds are generally contaminated by bacterial pathogens of all types. Wound cultures of established infections for aerobes and anaërobes should be taken and the patient given anti-tetanus prophylaxis. First aid and definitive treatment are similar to those given for tissue loss, lacerations, etc (see page 25).

SEALS AND SEA LIONS

Class MAMMALIA
Order PENNIPEDIA

Generally seals and sea lions are not aggressive toward humans, but in certain circumstances attacks may occur. The following are representatives of genera which should be regarded as dangerous.

Family OTARIIDAE

This family of six genera and 13 species has small external ears, and ranges from arctic and temperate to subtropical waters. The family includes both sea lions and eared seals.

***Zalophus californianus* (Lesson)**
California sea lion

DESCRIPTION Large males may reach 2.3 m (7.5 feet) and weigh 280 kg (620 pounds) but females are much smaller; grayish or brownish above, lighter below, often spotted or flecked.

HABITAT Rocky coastal areas of the west coast of North America and the Galapagos Islands.

OTHER POINTS Often displayed as the trained seal of the circus or zoo. There are several sub-species based on their distributional range.

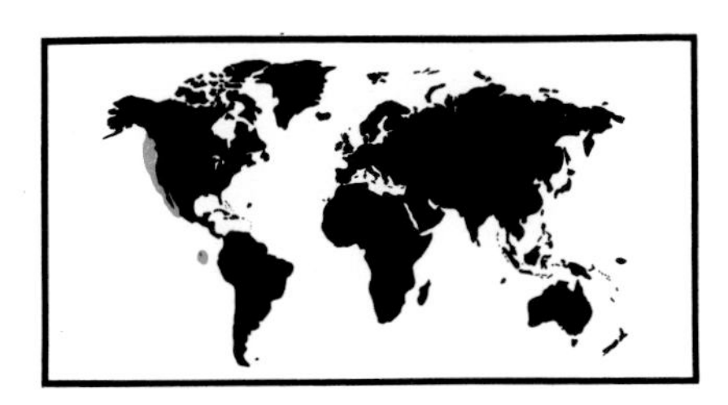

Family PHOCIDAE

This family of 13 genera and 18 species inhabits coastal waters in polar, temperate and tropical regions but is most numerous in colder areas. Members of this family are known as the true, earless, or hair seals and elephant seals.

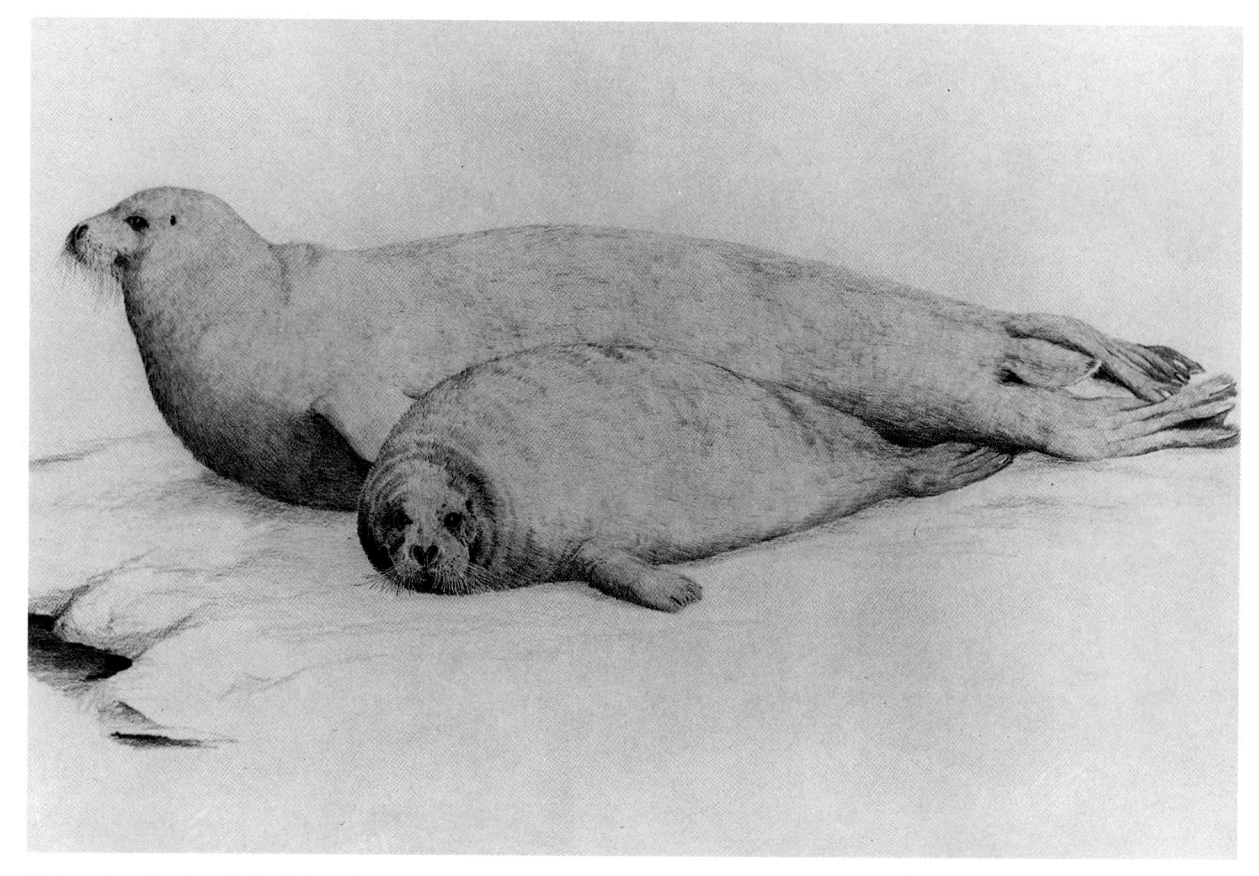

***Erignathus barbatus* (Erxleben)**
Bearded seal

DESCRIPTION Length up to 2.7 m (9 feet).

HABITAT Along coasts and on islands or ice-flows not far from land. Inhabits the edge of the ice along the coasts and islands of North America and northern Eurasia.

OTHER POINTS Have been known to inflict serious bites, especially in the mating season when they become aggressive. Generally solitary creatures.

SEA LION AND SEAL ATTACKS

Prevention

Sea lions and seals are usually docile, but the bulls may become aggressive during the mating season (October to December). Treat these creatures with respect and keep at a distance. This is especially important in relation to female sea lions or seals that have males or young in attendance.

Treatment

Sea lion or seal bites are similar to dog bites in the pattern of wounds which they produce. Wounds may be contaminated by bacterial pathogens. First aid measures and treatment are similar to those employed for shark bites (see page 25), although the wounds are rarely as severe.

WHALES

Order CETACEA
Family DELPHINIDAE

This family consists of 18 genera and approximately 62 species which live in all the oceans of the world, and the estuaries of many large rivers.

***Orcinus orca* (Linnaeus)**
Killer whale

DESCRIPTION Length up to 9 m (30 feet); distinctive black and white coloration with white patch over eye; powerful jaws with cone-shaped teeth directed toward the throat for grasping and holding food.

HABITAT All oceans, ranging from tropical to polar latitudes.

OTHER POINTS Travelling in pods of up to 40 individuals, they are fast swimmers feeding on a variety of marine organisms including invertebrates, fish, seals, walrus, other whales and sea birds.

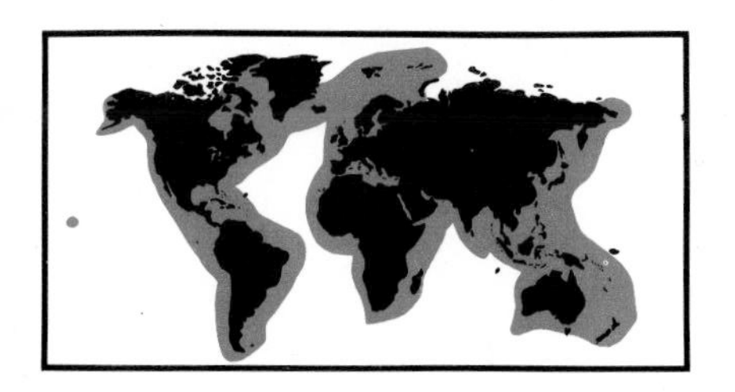

WHALE ATTACKS

Prevention
The ferociousness of the killer whale is a subject of debate as no reliably documented attack has been recorded. Nevertheless, this creature is equipped with an impressive set of 10 to 14 interlocking conical teeth on either side of the jaw. The crushing power of these jaws is enormous and a killer whale can snap a seal or a small porpoise in half with a single bite. Killer whales should be treated with respect and contact with them in the wild should be avoided. Although their intelligence has made them popular in oceanariums, there is always a chance that, while believed by experts not to seek humans for food, they could mistake a swimmer or diver for their regular prey.

Treatment
In the event of any attack, treatment should be as for shark bites (see page 25).

POLAR BEARS

Order CARNIVORA
Family URSIDAE

This family comprises seven genera with nine species. Females and young are potentially dangerous to man, although *Thalarctos maritimus* is the only species that is closely associated with the marine environment.

***Thalarctos maritimus* (Phipps)**
Polar bear

DESCRIPTION Length up to 2.5 m (8 feet), weight up to 750 kg (1650 pounds); creamy-white fur, hairy soles to feet and relatively longer legs than other members of the family.

HABITAT Throughout the Arctic.

BEAR ATTACKS

Polar bears are attracted by human activity and stores of food. Their natural curiosity makes them potentially dangerous to humans. In the event of any attack, treatment should follow that for shark bites (see page 25).

GIANT CLAMS

Phylum MOLLUSCA
Class PELECOPODA
Order EULAMELLIBRANCHIA
Family TRIDACNIDAE

This family comprises approximately seven to nine genera with about 20 species.

Tridacna gigas **(Linnaeus)**
Giant tridacna clam

DESCRIPTION Large bivalve shell with curved, waved edges; weight up to several hundred pounds.

HABITAT Abound in tropical waters and coral reefs of the Indo-Pacific.

OTHER POINTS Drownings are said to have occurred when divers accidentally stepped into the open shell, which closes in response to the intrusion, trapping the victim.

ACCIDENTS INVOLVING CLAMS

Prevention

Although accidents are rare, swimmers and divers should learn to recognize giant clams and avoid catching a foot or hand between the two valves of the clam.

Treatment

To release a person who becomes trapped, a knife must be inserted between the valves of the clam in order to sever its adductor muscles. Lacerations caused by the sharp calcified edges may bleed profusely. These should be vigorously cleaned and observed closely for delayed infections.

CHAPTER 2

VENOMOUS INVERTEBRATES

Many marine invertebrates contain venom which is used either as a defensive mechanism against possible predators or as a means of immobilizing prey prior to consuming them. The effects of these venomous species on humans varies considerably and can range from stings, which may be no more than mildly irritating, to reactions of extreme pain, which may prove fatal.

There are five main phyla which contain venomous marine invertebrates: Porifera (sponges); Coelenterata (sea anemones, hydroids, corals and jellyfishes); Mollusca (marine snails and octopuses); Annelida (marine bristleworms); and Echinodermata (sea urchins).

SPONGES

Phylum PORIFERA
Class DEMOSPONGIAE

Regarded as plants for many centuries, sponges were classified as animals by zoologists in 1835. For centuries, sponge fishing was a traditional coastal industry, particularly in the Mediterranean where it has been carried on from early civilizations to modern times. Only with the advent of man-made fibres and neoprene sponges has there been any decline in the use of natural sponges; even today, sponge fishing is a thriving industry.

Sponges are primitive multicellular animals that are found from shallow intertidal beaches to the deepest oceanic abyss. The soft cells of the body are supported by an internal skeleton generally consisting of tiny needle-like spicules. The identification of sponges is based upon the shape and composition of these spicules, which are made of silica (calcium carbonate) or a fibrous substance called spongin. Some sponges contain networks of spongin which remain intact once the outer living cells are removed; these form familiar bath sponges. There are about 4000 kinds of sponges, of which most live in the sea (only about 1% live in fresh water).

Sponges are sedentary and attach themselves to any suitable solid substrate, either shell or rock. Some burrow into shells or calcareous rocks. While most are small, some may reach sizes larger than a man and weigh hundreds of kilograms. Their body form is quite variable, but all sponges depend on a porous body surface through which water is filtered, allowing micro-organisms and organic debris to be harvested as food by specially modified cells. Like many sedentary organisms, sponges often become an integral part of other animals and themselves may provide a safe haven for small marine creatures. Thus, marine worms, crabs and shrimps, barnacles and blue-green algae are often found associated with them. Sponges are not preyed upon by many organisms, with the notable exception of the dorid nudibranch mollusks, which have specially modified alimentary tracts to handle the spicules which are a deterrent to most other species.

While most sponges are harmless to man, some species can cause a skin rash when handled, either as a result of the spicules or irritant substances in the body mucus.

SPONGE TRAUMA

Prevention
Swimmers or persons likely to come into contact with sponges are advised to wear suitable gloves when handling living specimens of the dangerous species.

Symptoms
During handling, the skin is exposed to chemical skin irritants, which may lead to a painful allergic-type contact dermatitis or an irritant spicule dermatitis. It is usually difficult to separate the effects of the spicules from allergic hypersensitivity.

Treatment
Following the development of a skin reaction after handling sponges, the skin should be gently dried and a layer of adhesive tape applied in an attempt to peel away superficial spicules. The affected area can then be bathed in 5% acetic acid (vinegar) for 15 to 30 minutes, three or four times daily. Alternatively, the area can be bathed in 40 to 70% isopropyl alcohol. For further medical advice see Chapter 7.

Family DESMACIDONIDAE

Microciona prolifera (Ellis and Solander)
Red moss sponge, redbeard

DESCRIPTION Young colonies grow on rock or shell surfaces; larger colonies become bush-like with branched lobes that may reach 20 cm (8 inches) high and 25 cm (10 inches) in diameter; color predominantly red, but may be orange-brown.

HABITAT Sometimes abundant on oyster shells, present on rocks in mud, on pilings in calm waters, and low intertidal zones. Atlantic coast of the USA from Cape Cod to South Carolina.

OTHER POINTS Poisoning occurs quite often among oyster fishermen who handle this species.

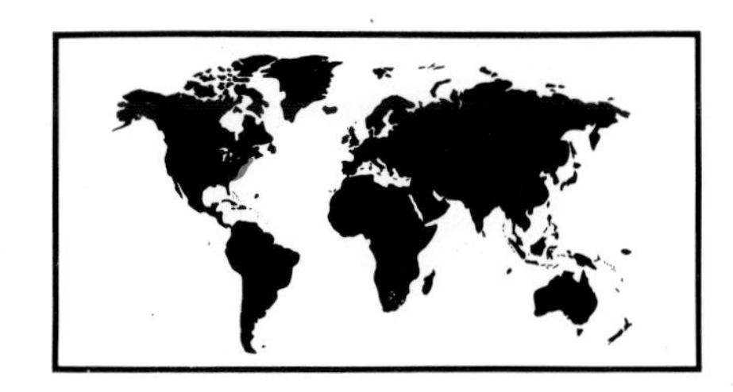

Family TEDANIIDAE

Tedania ignis (Duchassing and Michelotti)
Fire sponge

DESCRIPTION Encrusting sponge that may grow to 30 cm (12 inches) in diameter, but more usually about 10 cm (4 inches); typically deep red, but may also be pink to orange.

HABITAT On sides and open surfaces of low intertidal rocks, unattached on sediment, or attached to rocky substrate or other sponges. Occasionally found thickly encrusted on mangrove roots. Inhabits the West Indies.

OTHER POINTS The consistency is compressible, slightly resilient and easily torn. Known to cause skin irritation when handled.

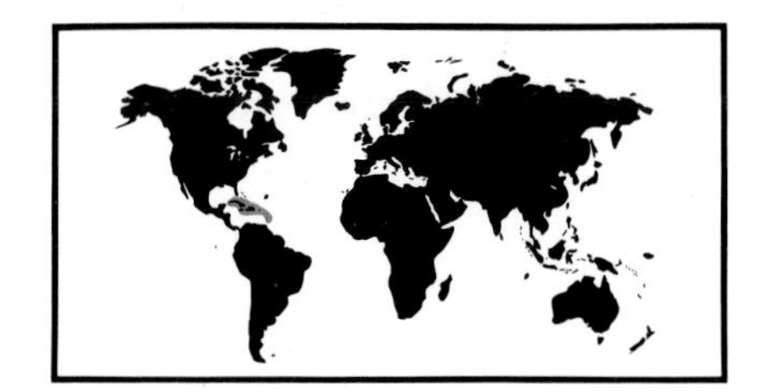

COELENTERATES

Phylum COELENTERATA
Class HYDROZOA
Family MILLEPORIDAE

Coelenterates, which include the hydroids, sea anemones, corals and jellyfishes, are simple multicellular animals in which the body is arranged with radial symmetry. Within the body wall is a cavity where food is digested. Food is normally caught on tentacles armed with batteries of stinging cells called nematocysts. Even common species not normally considered dangerous can produce an irritating rash on sensitive parts of the skin (ie, on the face, lips, or the underside of arms). Swimmers and skindivers are particularly susceptible to contact with anemones, corals and jellyfish.

HYDROIDS

The hydroids have a complex life history containing both sessile, polyp forms and the jellyfish-like free-swimming medusoid dispersal stages. The polyps often occur in feathery clusters on rocks and seaweeds, although some may develop into hard coral-like forms (*Millepora*), which develop to considerable size. The siphonophores show extreme intricacy in the development of body form. Unlike most hydrozoans, the predominant stage is pelagic, drifting on the sea surface with a gas-filled float and capturing food organisms with long-trailing tentacles containing stinging nematocysts. This group contains the infamous Portuguese man of war (*Physalia*), which carries virulent poison that can endanger human life.

Recent estimates put the number of species of hydroids at more than 2700, most of which are harmless, although a small number can inflict injuries on humans. The most common are listed below.

Diver exploring ***Millepora*** *in the Red Sea*

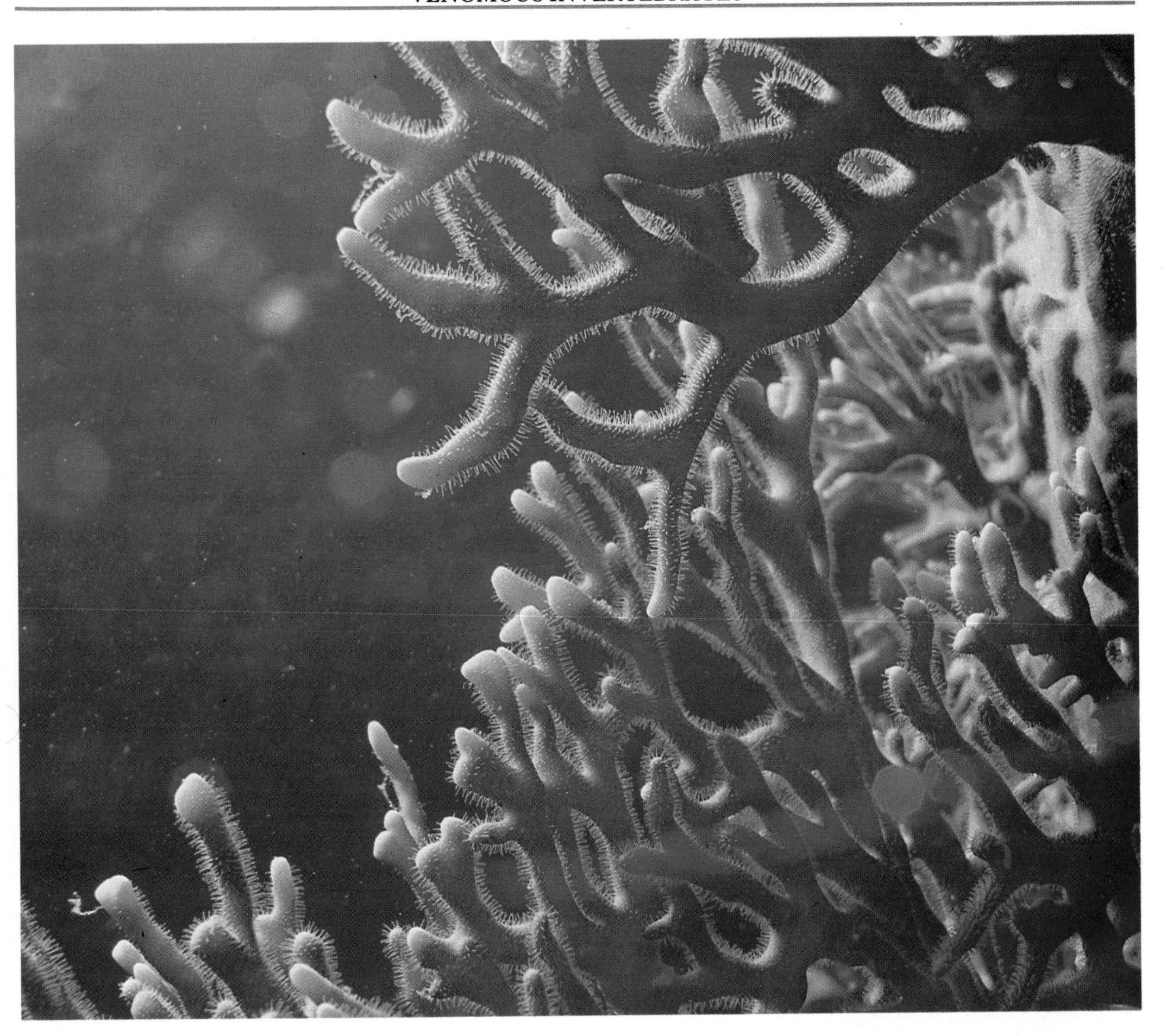

Millepora alcicornis **Linnaeus**
Stinging or fire coral

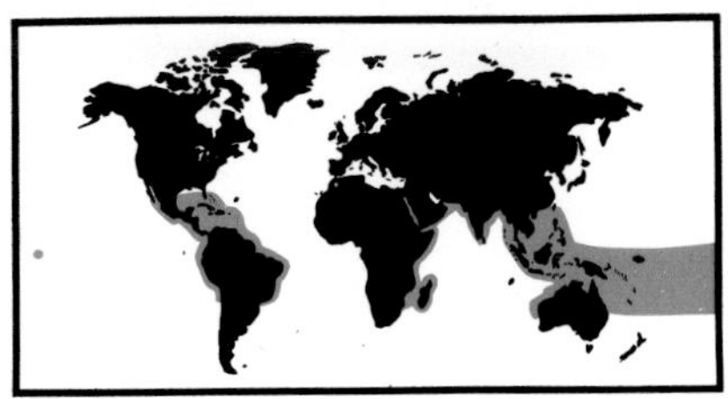

DESCRIPTION Size varies from a few centimeters to 50 cm (20 inches); colors range from white to yellow-green or brown. The family has a massive exoskeleton of calcium carbonate, and its surface is covered with numerous minute pores. Often not easy to recognize; may conform to the appearance of dead coral, sea fans and other objects it encrusts. Millepores have nematocysts which vary in stinging intensity according to species.

HABITAT "False" coral, generally found among true corals along reefs in the warm waters of the tropical Pacific and Indian Oceans, the Red Sea and the Caribbean. Branched form of *M. alcicornis* can occur at relatively great depth.

OTHER POINTS Can cause a painful skin rash. Hydroid corals are important in the development of reefs, forming upright, clavate, blade-like or branching calcareous growths, or encrustations over corals and other objects.

Family PHYSALIIDAE

This family contains only the Portuguese man of war. *Physalia*, the sole siphonophore with a unisexual colony, is distinguished by a contractile, horizontal, and gas-filled float.

Physalia physalis **(Linnaeus)**
Portuguese man of war

DESCRIPTION Float size of 10 to 30 cm (4 to 12 inches) in length. Commonly mistaken for a true jellyfish, it is a colonial hydroid. Depends largely upon currents, wind and tides for movement. The gas-filled float is an inverted, modified medusan bell. The number of fishing tentacles varies, depending on species. The Pacific form, *P. utriculus*, usually has a single fishing tentacle; the Atlantic species, *P. physalis*, has several. Tentacles may be 30 m (100 feet) long and are transparent. In one extended length of tentacle measuring 9 m (30 feet), scientists counted about 750,000 nematocysts.

HABITAT Almost always found floating at the surface of the water. Widely distributed as a group, but most abundant in warm waters. *P. physalis* lives in the tropical Atlantic, north to the Bay of Fundy, the Hebrides, Caribbean, and the Mediterranean Sea. The smaller *P. utriculus* is found in the Indo-Pacific area, Hawaii and southern Japan.

OTHER POINTS The loggerhead turtle, *Caretta caretta*, has been reported to feed on *Physalia*, despite the potency of the toxin and the ability of the *Physalia* nematocysts to penetrate even a surgical glove. The nudibranchs *Glaucus* and *Glaucilla* may feed on *Physalia*, consuming and storing the nematocysts to use them later in self-defense. *Physalia* attacks may be lethal.

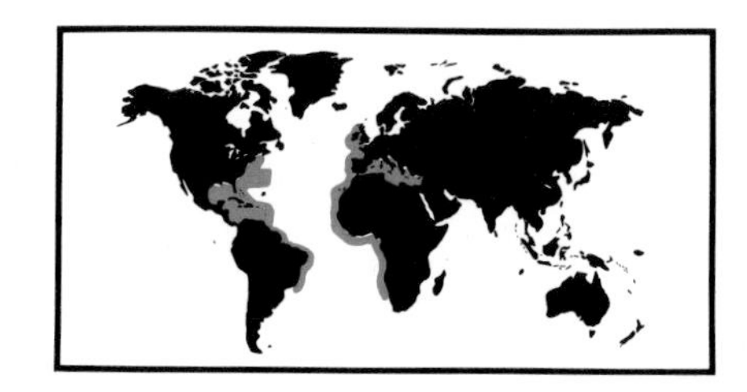

Close-up of the fishing tentacles of the Portuguese man of war, ***Physalia physalis****, which contain batteries of stinging nematocysts*

Family PLUMULARIIDAE

This is one of the most advanced families of leptomedusan hydrozoans; it contains about 17 genera. These hydroid colonies do not have a medusa stage.

Aglaophenia cupresina **Lamouroux**
Stinging hydroid, stinging seaweed

DESCRIPTION Colony may reach a height of 12.4 cm (5 inches); light brown in color. Looks very much like seaweed, hence is sometimes mistakenly called stinging seaweed.

HABITAT Grows on rocks, seaweeds and pilings. Tropical Indo-Pacific including the Indian Ocean, from Zanzibar to north Australia and the Philippines.

OTHER POINTS Capable of causing nettle-like stings with red welts and small tense blisters that may last several days.

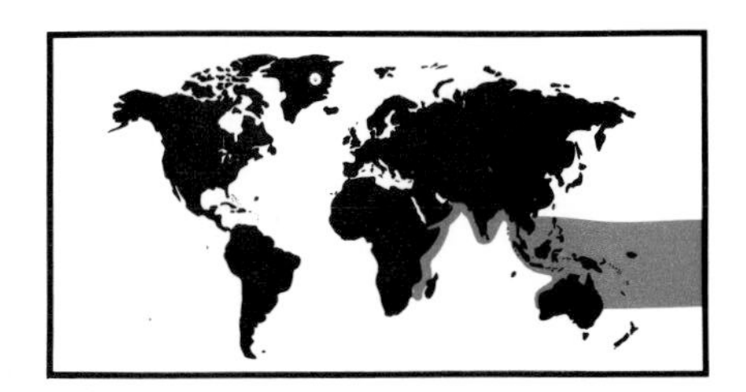

Lytocarpus nuttingi **Hargitt**
Stinging hydroid

DESCRIPTION Height of the colony may reach 20 cm (8 inches); feather-like organism, sometimes mistaken for seaweed.

HABITAT Attach to pilings, rafts, shells, rocks or algae. Found throughout the Indo-Pacific, with some closely related species in the West Indies.

OTHER POINTS All stinging hydroids are capable of inflicting painful and indolent stings.

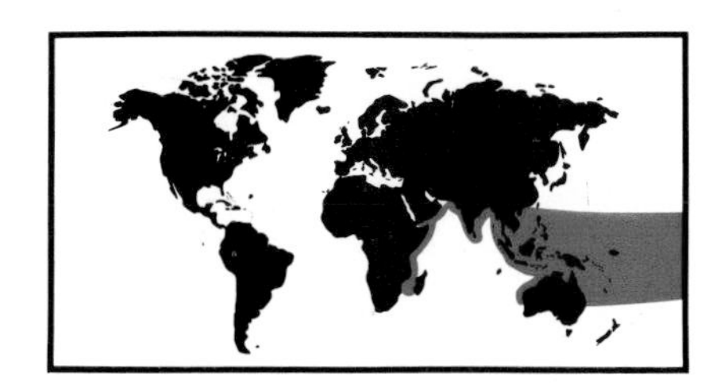

JELLYFISHES

Class SCYPHOZOA

Jellyfishes are free-swimming, pelagic coelenterates with radial symmetry which swim by regular contractions of the bell-shaped gelatinous body. Like many sea creatures, they go through seasonal "breeding" cycles, which means the risk from contact with the dangerous species can be reduced by respecting local proliferations. While all jellyfishes are capable of stinging, only a few are considered a major hazard. Of these, the Indo-Pacific box jellyfish, *Chironex*, is the most dangerous.

Family CHIRODROPIDAE

This family contains three genera: *Chirodropus, Chironex* and *Chiropsalmus*. Several species have been implicated in human fatalities. *Chironex* is among the most venomous of all marine creatures. These medusae are distinguished by four interradial clusters of tentacles and four stomach pouches with eight diverticula. It is important to remember that they are strong swimmers and have a powerful sting that can be fatal to humans, hence the common generic name "sea wasp".

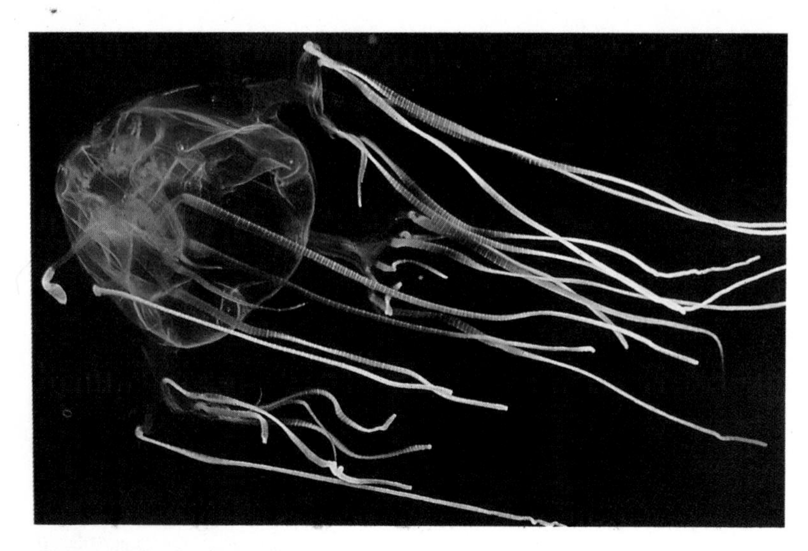

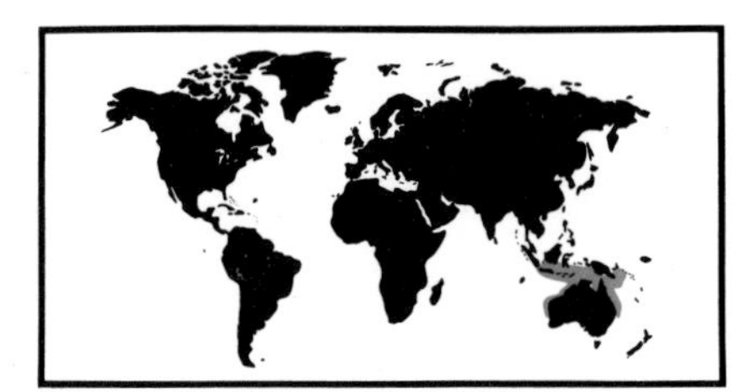

Chironex fleckeri **Southcott**
Box jellyfish, sea wasp

DESCRIPTION Body is generally a lucid color with hollow tentacles tapering to blunt points. The tentacles are highly contractile and may extend more than 1 m (39 inches). The bell may reach 11 cm (4.5 inches) in height. Graceful swimmer, capable of moving at a steady 2 knots. Bears a superficial resemblance to the hydromedusae in the shape of the box-like bell and in the presence of an annular diaphragm that constricts the aperture of the bell cavity.

HABITAT Found along the northeast coast of Australia. Prefers sandy bottoms and quiet shallow waters of protected bays and estuaries. In summer the immature forms, which stay on the bottom, reach maturity. The adults may then be found swimming at the surface.

OTHER POINTS The cause of numerous deaths along the north Queensland coast of Australia. It is capable of inflicting a serious sting on the skin, causing a severe rash associated with intense pain. People have been known to die within a few minutes of being stung as a result of circulatory and respiratory failure. It has a highly developed eye.

Chiropsalmus quadrigatus **Haeckel**
Sea wasp

DESCRIPTION Body height usually about 4.5 cm (1.75 inches). Under natural conditions the body is almost invisible under water. Five tentacles on each pedalium; the unpaired tentacle on each pedalium is usually tinged with lavender, while the first tentacles on each may be orange or yellow; the remainder are whitish.

HABITAT Seems to prefer sandy bottoms and the quiet shallow waters of protected bays and estuaries, although some species have been found in the open ocean. Light-sensitive, descending to deeper water during the bright sun of midday and coming to the surface during early morning, late afternoon and evening. Distributed in northern Australia, the Philippines and the Indian Ocean.

OTHER POINTS Sometimes confused with *Chironex fleckeri*, which it closely resembles. Stings may be very severe, but are generally less dangerous than those from *C. fleckeri*.

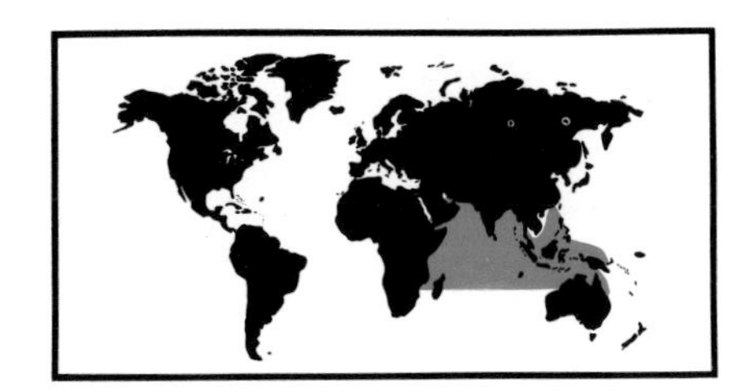

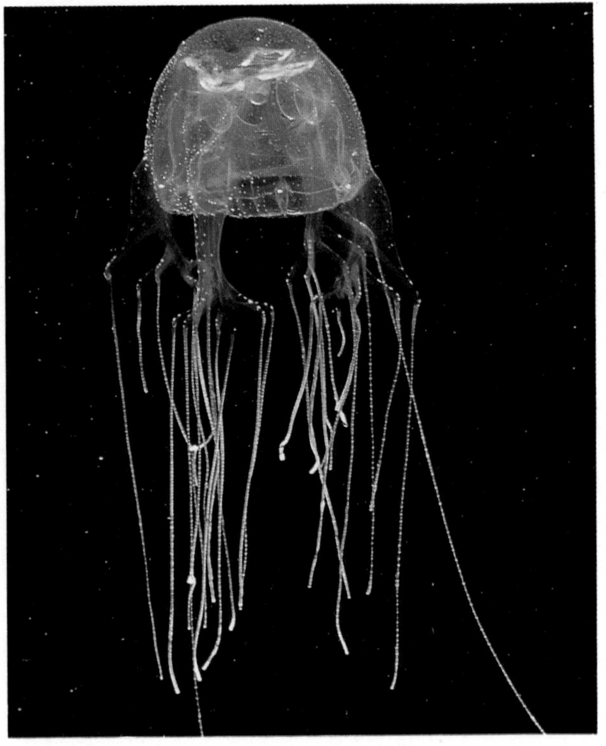

Chiropsalmus quadrumanus **(Müller)**
Sea wasp

DESCRIPTION Has 12 tentacles dangling from almost tranparent body. Closely related to *C. quadrigatus*.

HABITAT See *C. quadrigatus* and *Chironex*. Found in the Atlantic from North Carolina to Brazil, the Caribbean, the Indian Ocean and north Australia.

OTHER POINTS Stings are believed to be similar to those produced by *C. quadrigatus*.

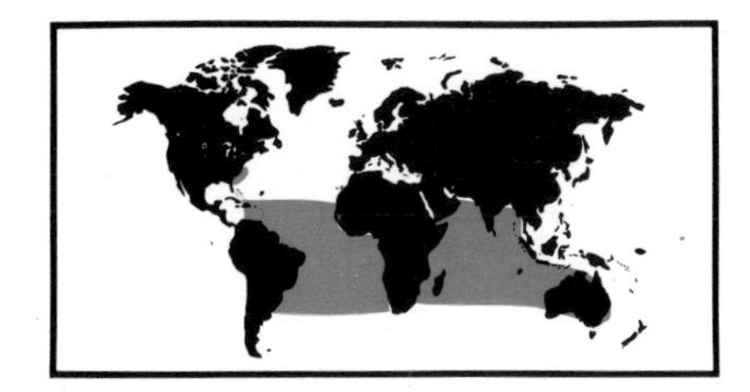

Family CARYBDEIDAE

This family contains three genera: *Carybdea*, *Tamoya* and *Tripedalia*.

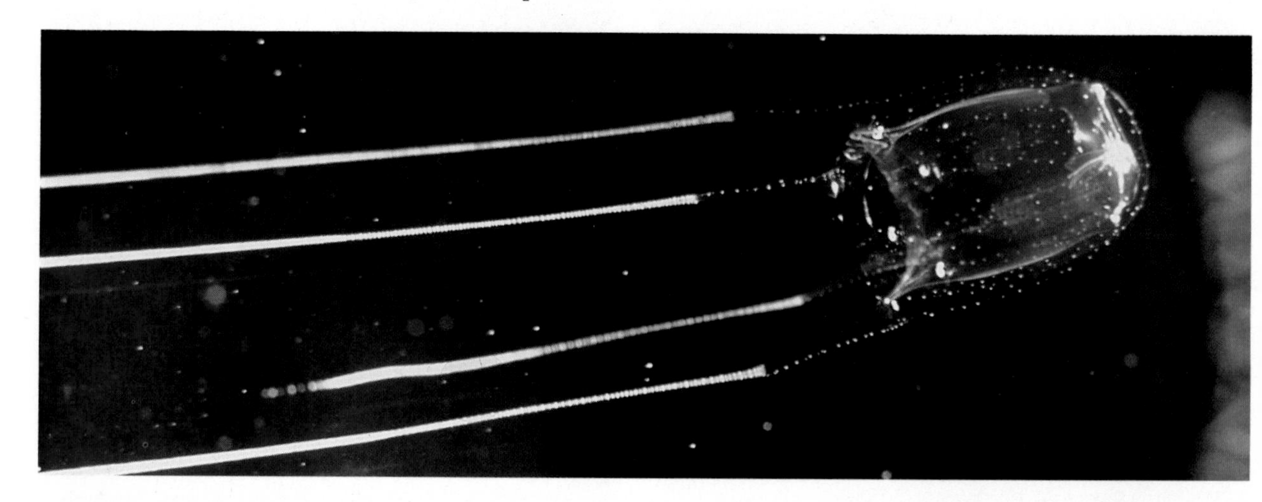

Carybdea alata Reynaud
Sea wasp

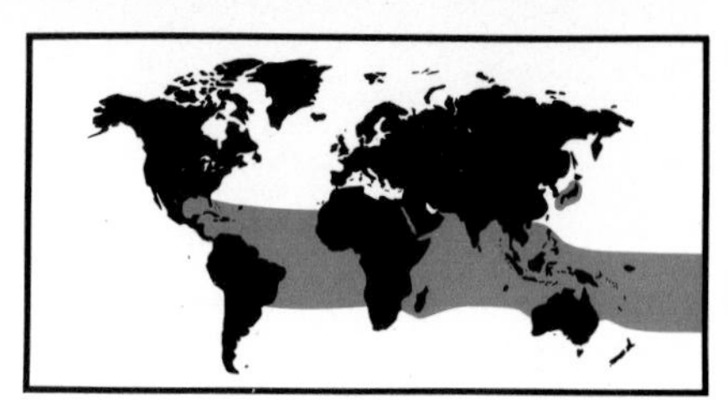

DESCRIPTION Bell can range from a few to several centimeters in length. Lucid and somewhat colorless appearance with trailing tentacles which can be contracted or lengthened. Like other cubomedusae, they are the fastest swimmers among the coelenterates.

HABITAT Prefers quiet shallow waters of protected bays and estuaries, and sandy bottoms; some species have been found in the open ocean. Found in shallow tropical seas throughout the Pacific, Atlantic and Indian Oceans. During summer, adults may be found swimming at the surface. They descend to deeper water at times of bright sunlight.

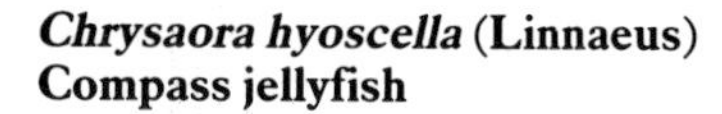

Chrysaora hyoscella (Linnaeus)
Compass jellyfish

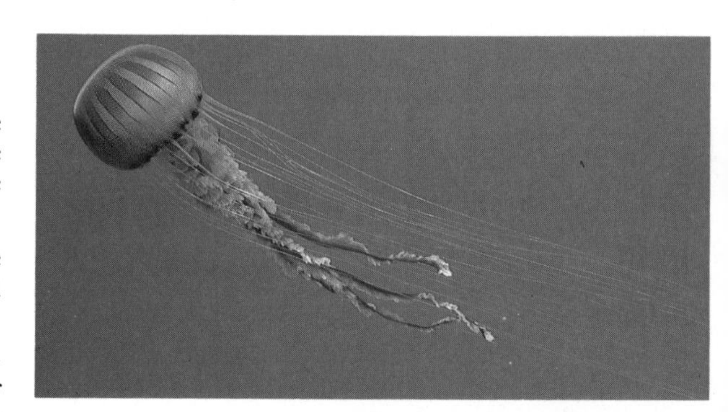

DESCRIPTION Highly variable coloring, sometimes pale yellow or with radiating streaks of red-brown lines on the flatly-rounded disk which has 24 trailing tentacles. The Mediterranean species often has more vivid coloration.

HABITAT The Atlantic coasts of Europe, the Mediterranean, Malay Archipelago, New Zealand and Japan.

OTHER POINTS The injury inflicted is generally considered to be in the moderate (rarely causing loss of consciousness) to severe class (which can cause loss of consciousness and possibly lead to drowning). Can inflict a painful sting.

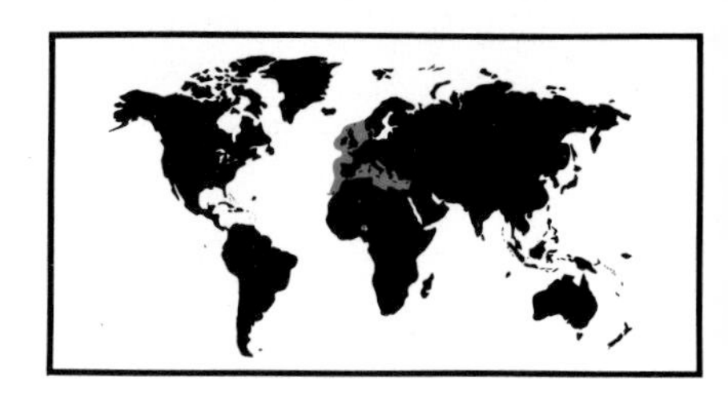

Family CYANEIDAE

Some of these medusae are very large, and are distinguished by a single mouth opening. The family contains three genera: *Cyanea*, with 13 nominal species; *Desmonema*, with two species; and *Drymonema*, also with two species.

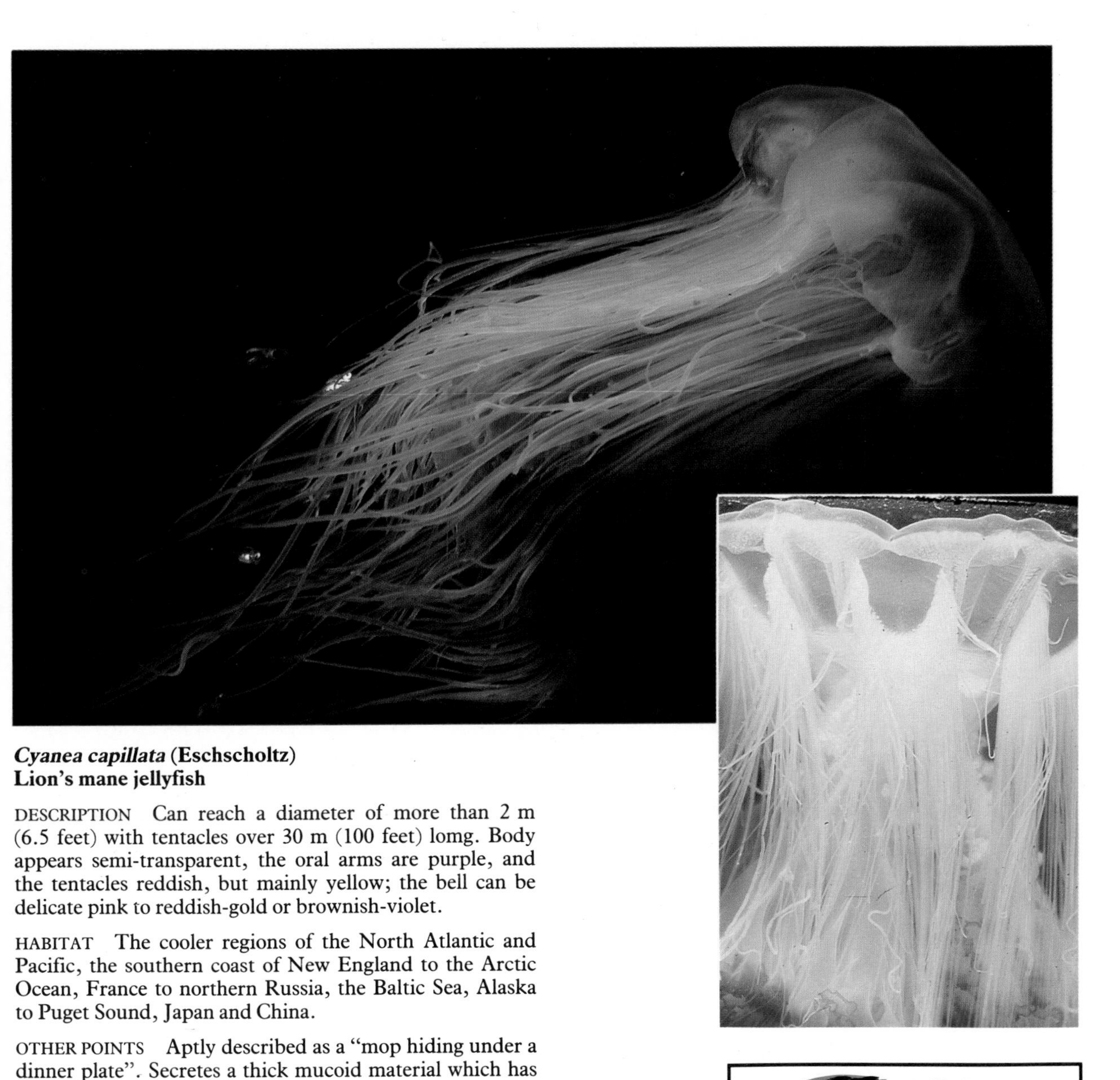

***Cyanea capillata* (Eschscholtz)**
Lion's mane jellyfish

DESCRIPTION Can reach a diameter of more than 2 m (6.5 feet) with tentacles over 30 m (100 feet) lomg. Body appears semi-transparent, the oral arms are purple, and the tentacles reddish, but mainly yellow; the bell can be delicate pink to reddish-gold or brownish-violet.

HABITAT The cooler regions of the North Atlantic and Pacific, the southern coast of New England to the Arctic Ocean, France to northern Russia, the Baltic Sea, Alaska to Puget Sound, Japan and China.

OTHER POINTS Aptly described as a "mop hiding under a dinner plate". Secretes a thick mucoid material which has a strong, fishy odor. Capable of inflicting a painful sting.

CORALS AND SEA ANEMONES

Class ANTHOZOA

The class Anthozoa is comprised of two orders: Alcyonaria, which includes soft corals, sea ferns, sea pens and sea pansies; and Zoantharia, which includes sea anemones and corals. All together these coelenterates number at least 6500 species. While the soft corals and some sea anemones may live at great depths, most occur in shallow surface and inshore waters. Most corals need to occupy shallow waters, as they are dependent on commensal symbiotic algae whose metabolic byproducts are a source of nourishment. Most anthozoans are harmless to man but a few have stinging cells dangerous to man and many corals. Though often extremely beautiful, corals are often fragile and razor sharp and can inflict severe wounds on persons who brush against them.

The following list contains a few of the more common species likely to be encountered, but all should be treated with respect.

Family ACROPORIDAE

This family contains four existing genera: *Acropora, Anacropora, Astrepora* and *Montipora*. *Acropora* is the major genus of the corals, with more than 200 species, making up about 40% of true corals. Species of *Acropora* are known for their roles as reef builders and depositers of calcium carbonate.

***Acropora palmata* (Lamarck)**
Elk horn coral

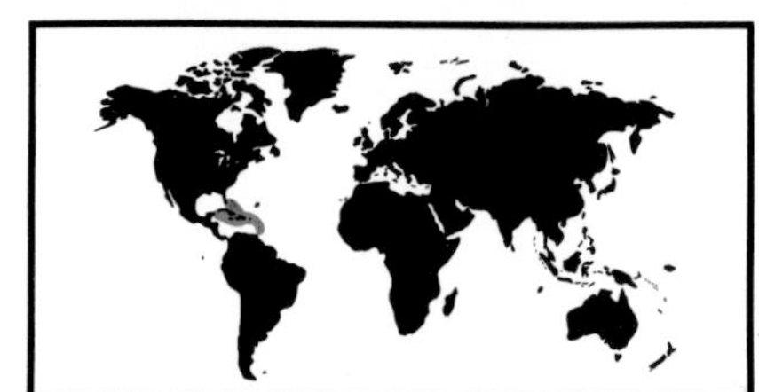

DESCRIPTION May attain several meters in diameter. Usually has short, thick, yellow-brown and flattened branches, hence the name, but in exposed conditions may be totally encrusted.

HABITAT One of the most abundant stony corals in shallow water, normally growing toward the incoming surf in areas of regular wave direction. Found in the Florida Keys, Bahamas and the West Indies.

OTHER POINTS It is razor sharp and can inflict serious wounds, which are slow to heal and may become secondarily infected.

Family ACTINIIDAE

The actinians include a great variety of species that reside mainly in the coastal zone, where they occupy shallow waters and the intertidal zone. Sea anemones have a simple, radial and bag-shaped body with a ring of tentacles around the rim. The tentacles are armed with nematocyst stinging cells.

Actinia equina Linnaeus
Beadlet anemone, strawberry anemone

DESCRIPTION When extended and fishing for water-borne food, may reach 5 cm (2 inches) and be 6 cm (2.3 inches) across the basal disk. Usually a deep red, but may vary considerably from reddish brown to limpid green in specimens without access to light. Its name comes from 24 pale blue beadlets located at the base of the ring of tentacles. The beadlets deter others of the species from impinging on the host's territory and have special stinging cells for this purpose. Like most sea anemones, they are sessile and capable of slow locomotion. The spotted variety is called the strawberry anemone.

HABITAT An intertidal species found on rocky shores, roughly at the low water mark, and most commonly behind rocks away from the pounding surf. Inhabits the eastern Atlantic from the Arctic Ocean to the Gulf of Guinea, the Mediterranean Sea, the Black Sea and the Sea of Azov.

OTHER POINTS Generally harmless but can inflict painful stings with accompanying erythema and swelling when in contact with sensitive parts of the body, such as beneath the arms, or on the face or lips.

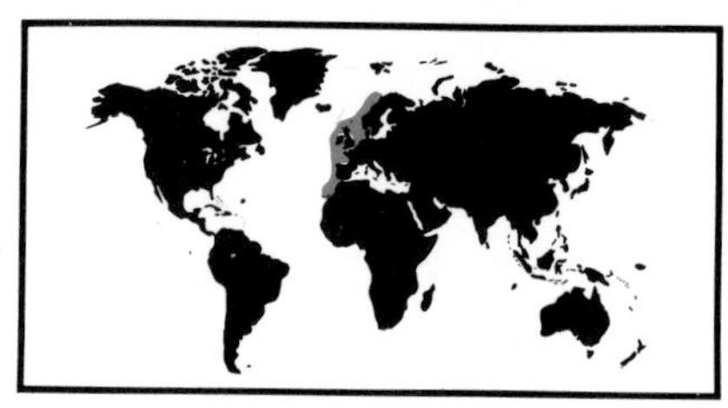

Anemonia sulcata (Pennant)
Snake locks anemone

DESCRIPTION Unlike most anemones it is non-contractile and unable to withdraw its tentacles into its basal body. The numerous tentacles are long (up to 10 cm or 4 inches) and yellow-brown or greenish, with violet-tinged tips. The tissues contain symbiotic algae which contribute to the anemone's color.

HABITAT Found in great numbers in intertidal pools and on rocky shores. In this position and exposed to the sun, they are characteristically green, but sub-littoral populations are normally dull brown. Inhabits the northeast Atlantic from Norway to the Mediterranean Sea and south to the Canary Islands.

OTHER POINTS Tentacles feel sticky, a sensation that results from the powerful nematocysts "stinging" the inquisitive finger. It is unusual to suffer a reaction; but on sensitive parts of the body, particularly the face and lips, the reaction can be severe with swelling, red wheals and itching.

Family ACTINODENDRONIDAE

This family comprises about 10 species in three genera: *Actinodendron, Actinostephanus* and *Megalactis*. The nematocysts are large and capable of inflicting painful stings upon humans. Despite this fact, these actinians are frequently eaten in parts of eastern Indonesia.

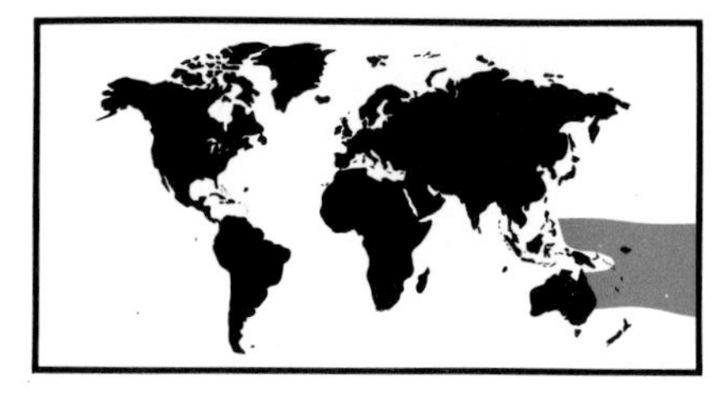

Actinodendron plumosum Haddon
Hell's fire sea anemone

DESCRIPTION Length varies from 15 to 30 cm (6 to 12 inches). When extended it takes on a flower-like appearance, but when in contracted state it has been described as appearing to have a "top hat". Varies in coloration from dull green to a lighter shade of green at its tips.

HABITAT Usually found on the shady side of rocks or under coral ledges, but never in the sand, in the tropical Pacific, including the areas of the Great Barrier Reef and Micronesia.

OTHER POINTS Can cause serious and very painful stings, similar to those produced by a sting in nettle but perhaps more severe, with the effects lasting for several months. Can cause skin ulcerations.

Family ALICIIDAE

This family of primarily tropical, shallow-water actinians is distinguished by having branching outgrowths from the column. The tentacles are long and may have spots of concentrated nematocysts.

Triactis producta **Klunzinger**
Sea anemone

DESCRIPTION Diameter of the base is about 4 cm (1.6 inches); when fully extended it can reach a height of 8 cm (3 inches). During the day the tentacles are contracted, and during the night they are extended.

HABITAT Intertidal, found primarily in the Red Sea.

OTHER POINTS One of the more dangerous sea anemones that can cause a very painful sting which increases in intensity and may later ulcerate.

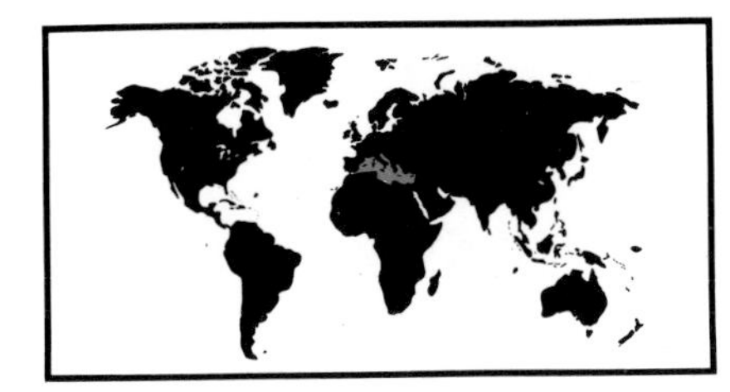

Family HORMATHIIDAE

The family contains 15 genera and about 100 species. Some species are intertidal, but many have been recorded from deep seas. Many occur on mollusk shells or envelop a ball of mud and stones with the pedal disk. Many members of the genus *Adamsia* occur symbiotically with crustaceans, being found on the cheliped or else attached to the shell of hermit crabs.

Adamsia palliata **(Bohadsch)**
Cloak anemone

DESCRIPTION Typically pale brown with deep red spots. As the name implies, it is found over and around shells used by the hermit crab *Eupagurus prideauxi*.

HABITAT Intertidal areas from Norway to Spain, and the Mediterranean Sea.

OTHER POINTS These anemones are widespread and, as with all other anemones, care must be taken to prevent the nematocyst stinging cells coming into contact with sensitive parts of the skin.

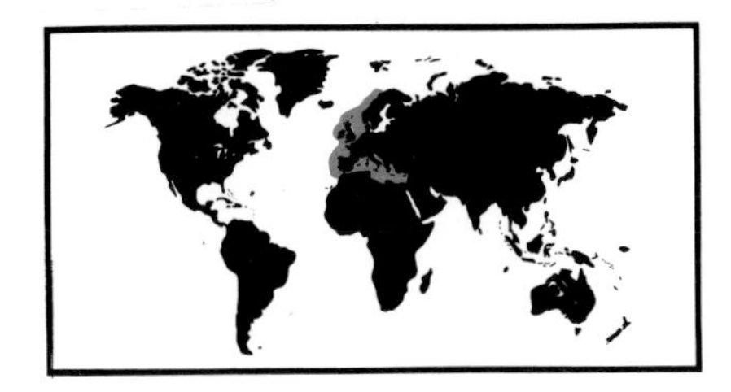

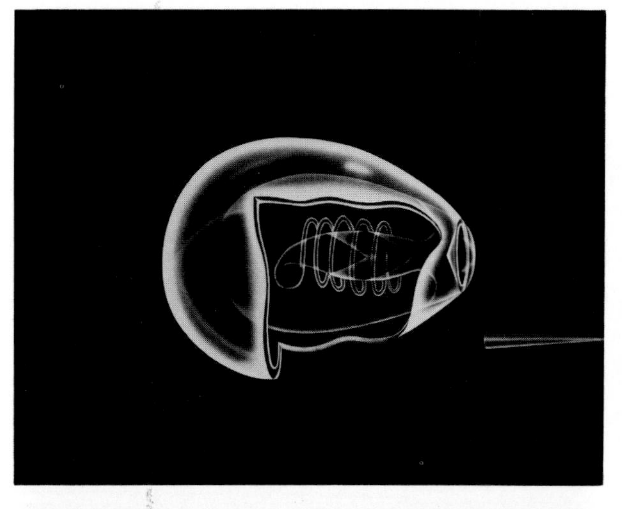

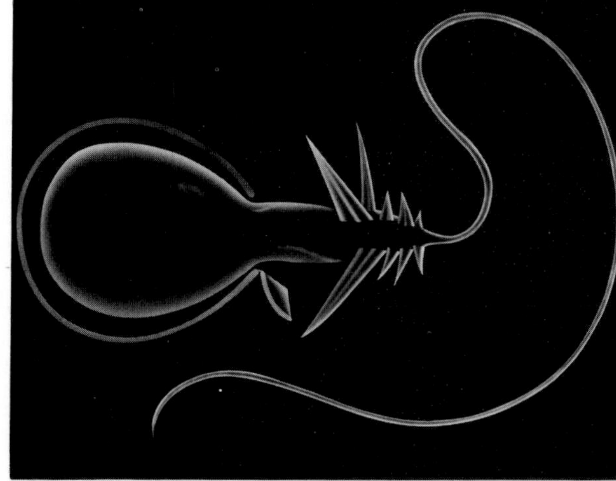

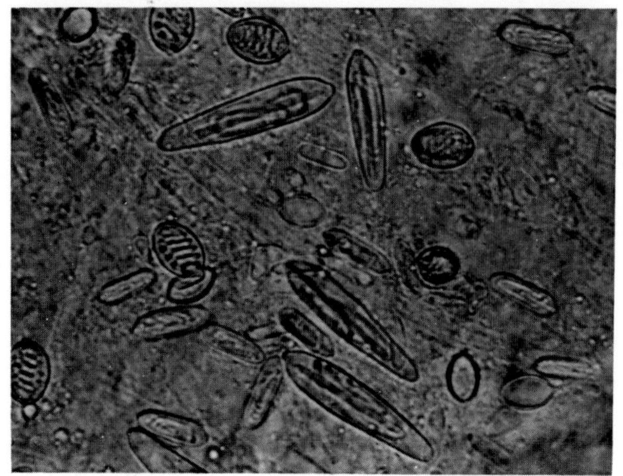

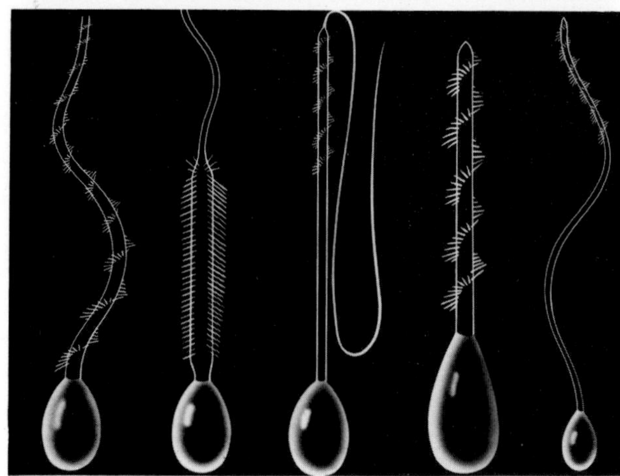

COELENTERATE STINGS

Venom apparatus of coelenterates

The coelenterates, including the hydroids, jellyfishes, sea anemones and corals, all possess tentacles equipped with stinging nematocysts which are located on the outer layer of the tentacles. The nematocyst is a small venom-filled capsule containing a hollow coiled thread which, when triggered, is everted and used to inject poison into the body of its prey. When human skin comes into contact with the tentacles, the nematocysts are triggered, causing the discharge of venom into the skin. To brush with an anemone is to trigger many thousands of skin injections, the severity of which depends upon the species touched and an individual's sensitivity.

Prevention

It is important to bear in mind that the tentacles of some species of jellyfish may trail as much as 30 m (100 feet) or more from the body. Consequently, jellyfishes should be given a wide berth. Rubber diving suits are useful in affording protection from stings. Even though appearing dead, jellyfishes or jellyfish fragments washed up on the beach may be quite capable of inflicting a serious sting. The tentacles of some jellyfish may adhere to the skin. Swimming soon after a storm in tropical waters in which large numbers of jellyfish have been present should be avoided, as this may result in multiple stings from remnants of damaged tentacles floating in the water.

Treatment

The treatment of coelenterate stings has generated much speculative and conflicting opinion and involves a wide range of detoxicants, some of which are actually ineffective and may even enhance envenomation. Recent clinical investigations in the USA and Australia, with waters inhabited by some of the world's most dangerous coelenterates, have provided valuable therapeutic guidelines.

The primary objectives of first aid measures are to

Opposite, clockwise from top left: Nematocyst coiled; nematocyst released; different types of nematocyst; photomicrograph of the nematocysts of ***Chironex fleckeri****, the box jellyfish*

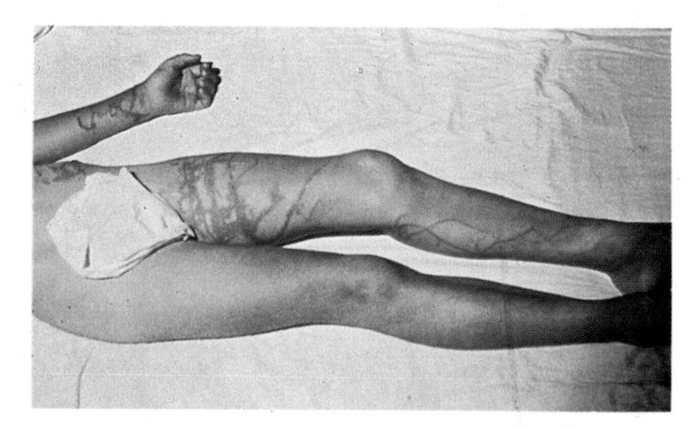

Effects of a ***Chironex*** *sting on the body*

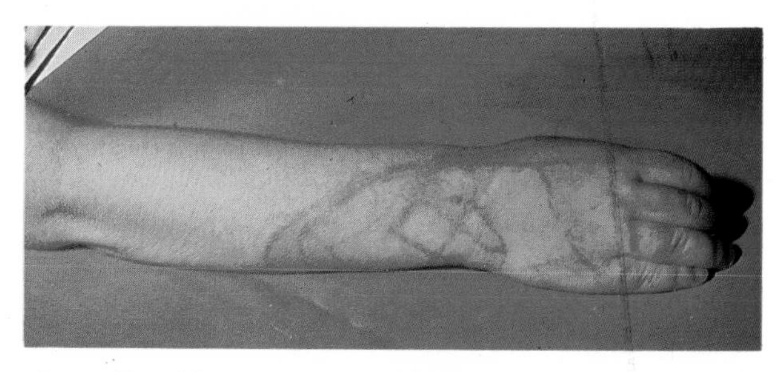

Arm affected by a sting from ***Chironex***

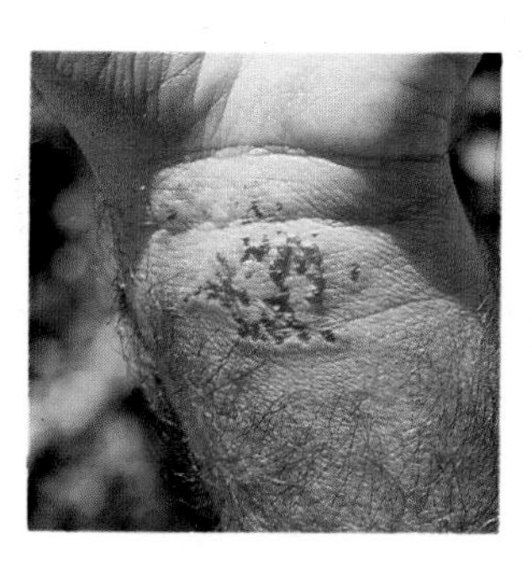

Wrist stung by the sea anemone ***Triactis producta***

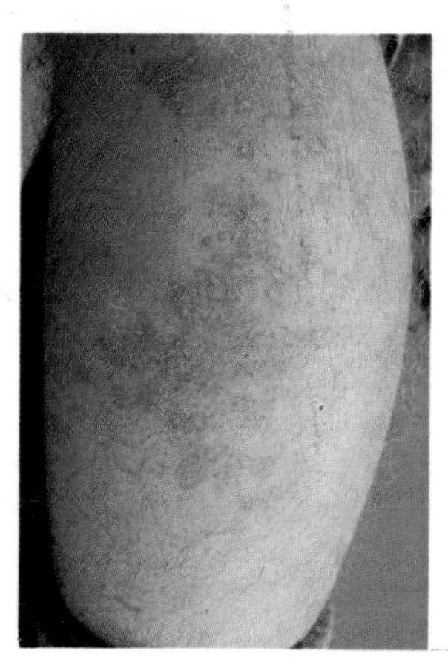

Actinodendron sting

minimize the number of nematocysts discharged into the skin and to reduce the immediate harmful effects that follow envenomation. The severity of the sting depends upon the nature of the venom, the number of nematocysts, and the size and health of the victim. Hydroid and anthozoan stings are usually the result of a sudden accidental brush against a relatively small number of nematocyst batteries. In the case of *Physalia* species and scyphozoan jellyfishes, there may be a prolonged encounter with tentacles carrying a vast number of nematocysts. Initial contact with the tentacles may result in only a modest envenomation. However, subsequent efforts to disengage the tentacles may result in further discharge of large numbers of nematocysts, which often worsens the envenomation. Thus, removal of the adherent tentacles must be done carefully, in order to minimize further envenomation.

If a person is stung, immediately rinse the wound with sea water, *not* with fresh water, because fresh water will stimulate any nematocysts that have not already discharged. Do not attempt to remove the tentacles with abrasive techniques. If it is absolutely necessary to remove the adherent tentacles, use the back edge of a knife blade, a shell or a stick. Rubbing the skin with a towel, rag, seaweed, or a handful of sand is to be avoided. The thickened skin of the palm of a rescuer's hand is usually impervious to a sting, but if at all possible, a rescuer should wear hand protection. Remove the tentacles with forceps if possible.

To prevent nematocyst discharge and to detoxify the venom, apply acetic acid 5% (vinegar) or isopropyl alcohol 40 to 70% to the skin. There is some evidence that alcohol may stimulate the discharge of nematocysts *in vitro*. The clinical significance of this fact is as yet unclear, as most clinical experience supports the effectiveness of alcohol. Methylated spirits, such as perfume and after-shave lotion, and high proof liquor (ethanol) should not be used, for in some cases they may prolong the agony. The detoxicant should be applied continuously for at least 30 minutes or until there is

Above: The coral ***Acropora cervicornis,*** *Virgin Islands*
Right: ***Acropora*** *corals, Australia*

marked pain relief. Other substances reputed to be effective as alternatives include dilute ammonium hydroxide, urine, sodium bicarbonate, and papain (papaya latex or unseasoned meat tenderizer). A proprietary product known as Stingose, an aqueous solution of aluminium sulphate 20% and surfactant 1.1%, is another alternative. Formalin is too toxic for skin application. The pressure immobilization technique for venom sequestration should not be used. A proximally placed venous-lymphatic constriction band should only be considered if a topical detoxicant is unavailable, the victim is suffering from a severe systemic reaction, and transport to medical care will be delayed.

Once the wound has been rinsed and soaked with acetic acid or alcohol, the remaining nematocysts must be removed. The easiest way to accomplish this is to apply shaving cream or a paste of baking soda, flour or talc and shave the area with a razor or similar sharp-edged object. If sophisticated tools are not available, the nematocysts may be removed by making a sand or mud paste with sea water and using this to help scrape the victim's skin with a sharp-edged shell or a piece of wood. The rescuer must take care not to become envenomed in the process; bare hands should be frequently rinsed. For further medical advice see Chapter 7.

CORAL CUTS

Prevention

Coral cuts represent an ever-present annoyance to the diver working in tropical areas. Despite their delicate appearance, stony corals have calcareous outer skeletons with razor-sharp edges that are capable of inflicting nasty wounds. When working in the vicinity of corals, every precaution should be taken to avoid contact with them. Corals should not be handled with bare hands. Leather or heavy cotton gloves, and rubber-soled canvas shoes or a completely soled flipper should be worn where possible.

Treatment

All fresh coral cuts and abrasions should be scrubbed thoroughly with soap and water and then irrigated vigorously with a forceful stream of fresh water or normal saline in order to remove all foreign material. If stinging is a major symptom, there may be an element of envenomation from nematocysts. A brief rinse with acetic acid 5% (vinegar) or isopropyl alcohol 40 to 70% usually diminishes the discomfort. It may be useful to use hydrogen peroxide one-half strength to bubble out small particles of "coral dust". For further medical advice see Chapter 7.

MOLLUSKS

Mollusks are unsegmented animals, with a distinct and well-developed head, ventral muscular foot and soft body often contained in a calcareous shell. The shell is secreted by skin tissues and overlays the body. The cavity may also contain the gills. The body form of mollusks is very variable. The group includes primitive forms, the chitons; the univalve snails, or gastropods, which are well represented by land, sea and freshwater snails and slugs; the bivalve mollusks, typically contained in a shell consisting of two valves; and the cephalopods, which include the octopus, squid and cuttlefish.

This large and successful phylum occupies most terrestrial and aquatic habitats, but only the cone shells and the cephalopods have been shown to contain venom harmful to man.

CONE SHELLS

Class GASTROPODA
Family CONIDAE

They have characteristic cone-shaped shells and, like other gastropods, a distinct head, "tentacles" with eyes, and a strong fleshy foot. The cone shell has a siphon tube to sample water (to detect prey), as well as a long proboscis to capture and seize prey. In this family the proboscis is variable in shape and carries a poison tooth or dart used to spear and immobilize small fishes and other items of food. The risk of being stung by the cone is of particular concern to swimmers and divers.

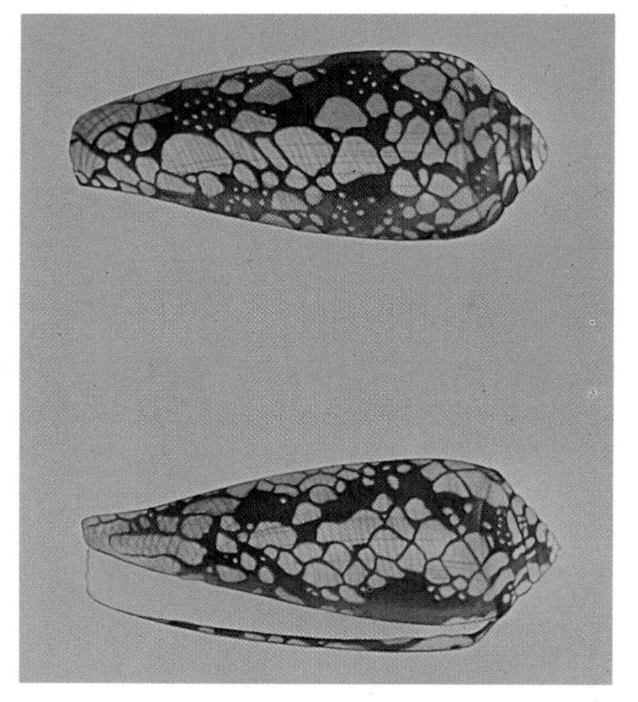

Conus aulicus **Linnaeus**
Court cone

DESCRIPTION Length from 7 to 15 cm (2.7 to 6 inches). Brown and white coloring but considerable variation in color and pattern. Normally reddish brown, but darker tones are not uncommon and pale gold tones are also found. Stands out from most other cones by the combination of large size, relatively slender shape and sharply-pointed spine.

HABITAT Found in sand and under dead corals in shallow water from Polynesia to the Indian Ocean.

OTHER POINTS One of the more dangerous species of cone shell, it is believed capable of inflicting a fatal sting.

Conus geographus **Linnaeus**
Geographic cone

DESCRIPTION Length 13 cm (5 inches). Very light in weight and thin in structure, yet solid and not fragile. Shell gloss is low or absent, color creamy white, pinkish, or bluish white. Heavily covered with a reticulated network of fine brownish lines. Reticulation may be distinct or very broken and mottled.

HABITAT Shallow water; widespread in the Indo-Pacific.

OTHER POINTS Probably the most dangerous species of cone shell. It has been involved in more fatal and serious human stings than has any other member of this group. The venom apparatus is well-developed and capable of delivering a relatively large quantity of venom.

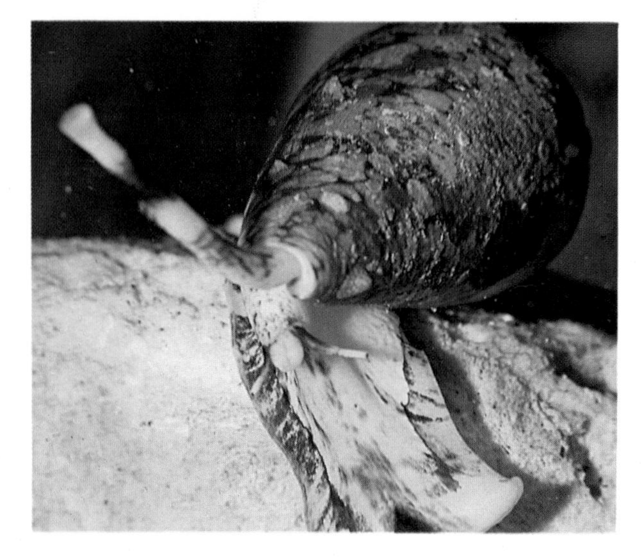

Conus marmoreus **Linnaeus**
Marbled cone

DESCRIPTION Length up to 10 cm (4 inches). Heavy, with a high gloss. The color patterns vary greatly.

HABITAT Shallow water; ranges from Polynesia to the Indian Ocean.

OTHER POINTS Has stung humans and may be capable of causing a fatality. The venom apparatus is large and well-developed.

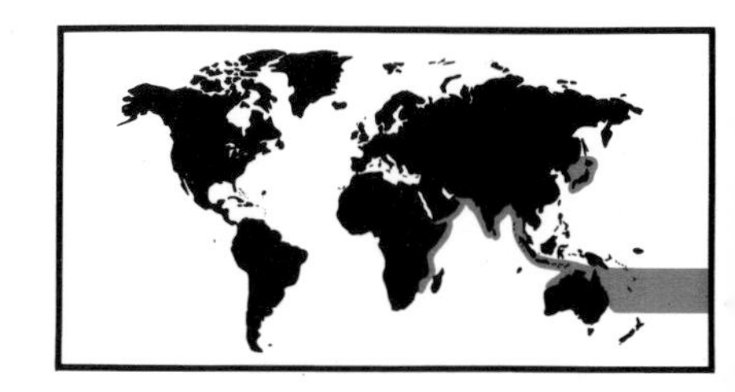

Conus striatus Linnaeus
Striated cone

DESCRIPTION Length 10 cm (4 inches). Moderately heavy, with a glossy sheen. Distinctive in its shape and the mottled pattern formed by numerous rows of narrow dark lines. Body color can be white, pink, bluish, brown, or even reddish, covered with variable darker mottlings forming two broad spiral bands above and below the midbody area.

HABITAT Shallow water; widespread and common across the Indo-Pacific from eastern Africa to Hawaii and French Polynesia.

OTHER POINTS Aggressive and dangerous species that has caused human fatalities. Has a large and well-developed venom apparatus.

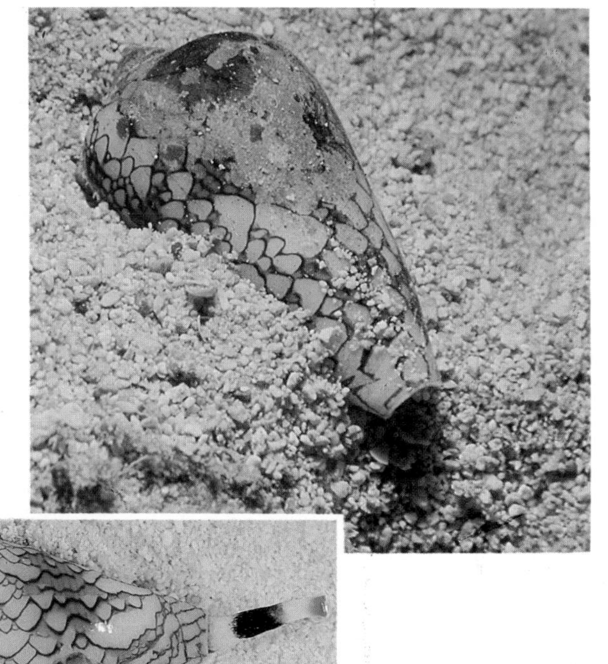

Conus textile Linnaeus
Textile cone

DESCRIPTION Length of up to 10 cm (4 inches). Of comparatively medium weight with a high gloss on the shell. The pattern is variable in color, size of blotches, length and darkness of "textile lines", and especially the size of the patterns, from rather small and crowded to large and open.

HABITAT Common in shallow water; widespread across the Indo-Pacific from eastern Africa to Hawaii and French Polynesia.

OTHER POINTS One of the more dangerous cone shells, having caused fatalities.

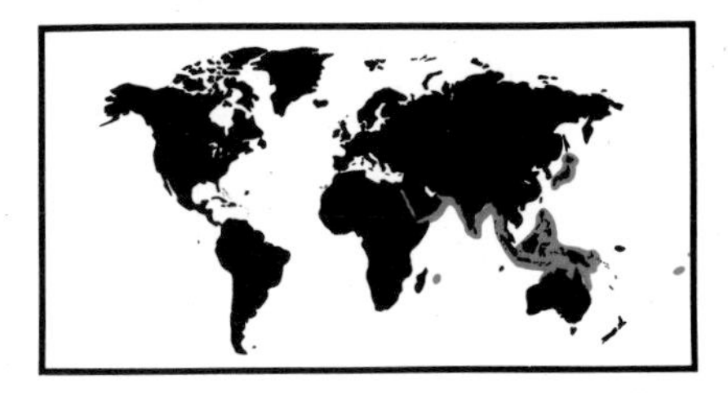

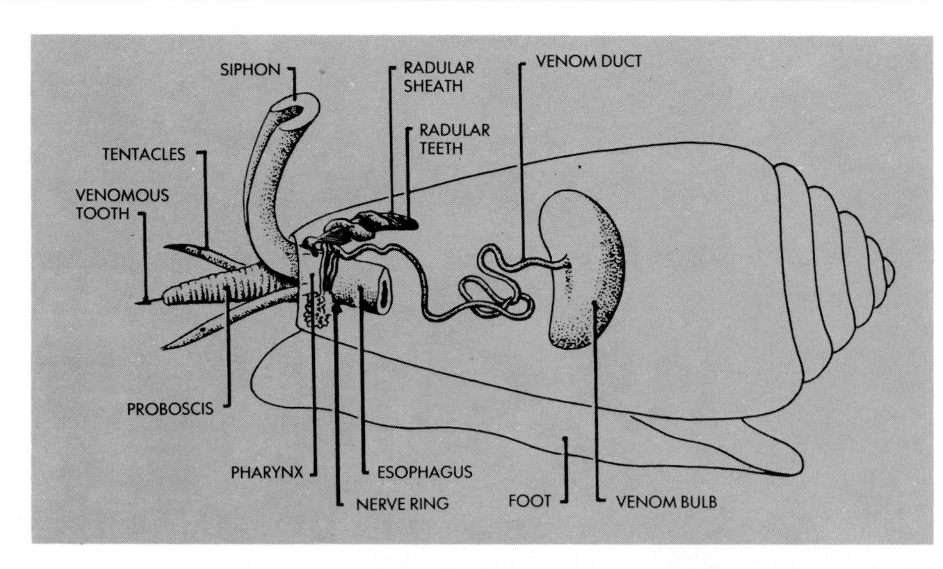

Drawing of the venom apparatus of a cone shell

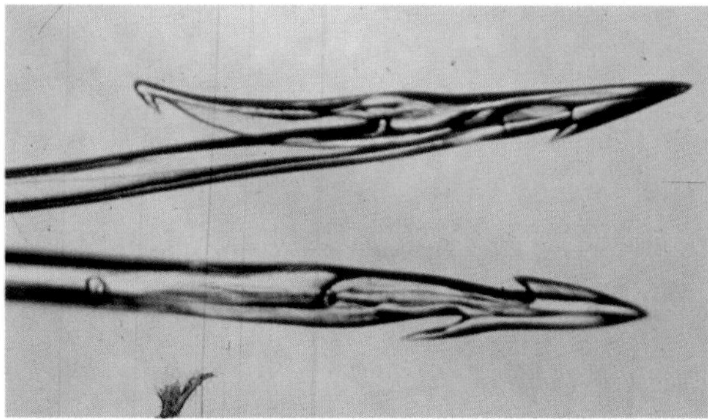

Radulae of ***Conus striatus***

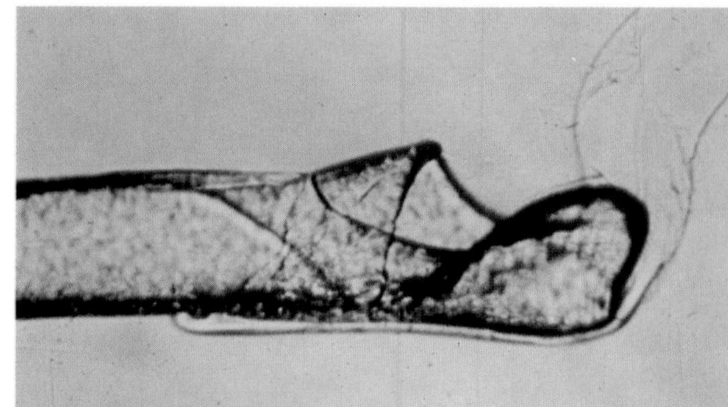

Base of ***Conus geographus*** *radular tooth showing the ligamentous attachment and opening through which the venom enters the hollow tooth. Length of shaft is 8 mm, diameter 0.2 mm*

CONE SHELL STINGS

Venom apparatus of cone shells

The venom apparatus consists of the venom bulb, venom duct, radular sheath, and radular teeth. It is believed that just before stinging, the radular teeth are released into the pharynx, then carried by the proboscis for thrusting into the flesh of the victim.

Prevention

Collectors should wear gloves and handle cones with care. Shells should be picked up by the large posterior end. Contact with the soft parts of the animal should be avoided. Cone shells should not be placed in pockets, as they are capable of inflicting stings through clothing.

Treatment

A cone shell sting shculd be considered to be potentially as severe as a snake bite. Unfortunately, an antivenom is not yet available for envenomation by a cone shell. Incision and suction are of little value and should be avoided. A new treatment method that is gaining popularity is the pressure-immobilization technique of venom containment, which can be carried out as follows. If practical, by virtue of the location of the sting, a cloth or gauze pad of approximate dimensions 6-8 cm x 6-8 cm x 2-3 cm should be placed directly over the sting and held firmly in place by a circumferential bandage 15 to 18 cm (6 to 7 inches) wide applied at lymphatic-venous occlusive pressure. The arterial circulation should not be occluded, as determined by the presence of distal arterial pulsations and proper capillary refill. The bandage should be released after the victim has been brought to proper medical attention and the rescuer is prepared to provide systemic support. For further medical advice see Chapter 7.

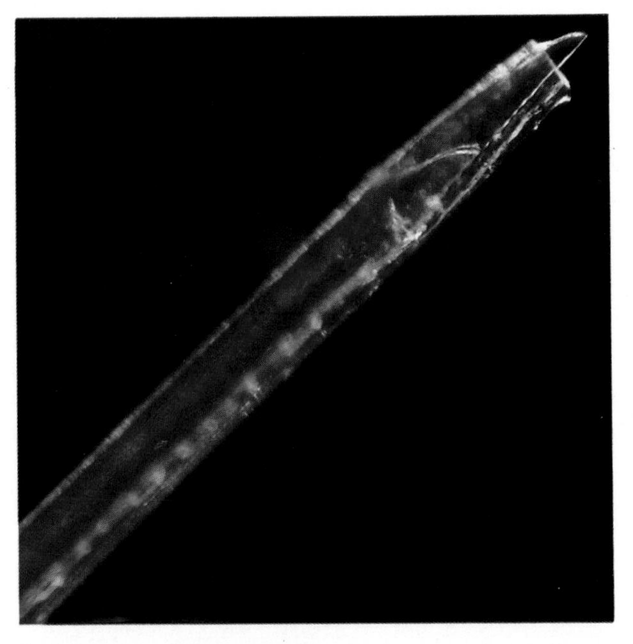

Radular tooth of ***Conus textile***

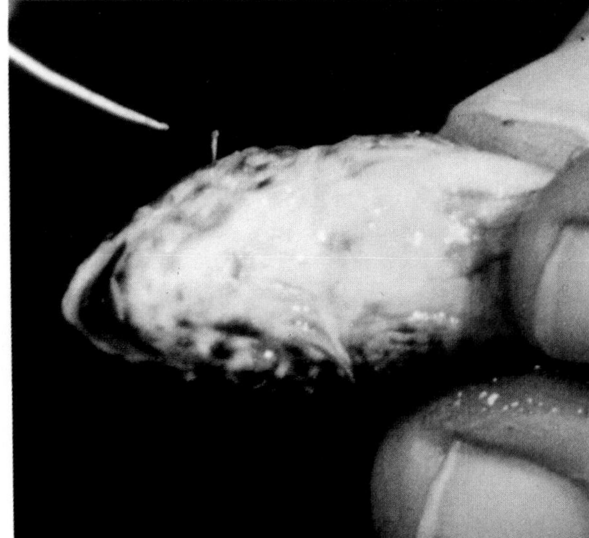

Fish stung by ***Conus***

Conus geographus *eating a fish*

CEPHALOPODS

This class includes the nautilus, squid, cuttlefish and octopus. The head is clearly defined and contains a well-developed brain, buccal mass containing a parrot-like horny beak, and well-developed eyes. The mouth parts are surrounded by 8 or 10 tentacles, armed with suckers and in some cases hooks, which are used to capture prey and in courtship, mating and egg care. The body is generally elongated and contains the visceral organs, which open into the mantle cavity where the gills are found. Swimming is effected either by undulating fins or by contracting the mantle, causing a jet of water to be shot from a short siphon. Using "jet propulsion", squids are capable of rapid movement.

While the nautilus has an external shell, the cuttlefish has an internal buoyancy chamber and the squid a much reduced "pen". The octopus has no skeleton and crawls on its tentacles, although it is capable of swimming using jet propulsion.

Few marine creatures have received greater attention from fiction writers than has the octopus. The result is that this remarkable, shy and intelligent creature is greatly over-rated as a hazard to swimmers. This so-called "demon of the depths" is generally small and retiring in habit, and certainly does not deserve its reputation. The largest octopus may extend to 10 m (32 feet) across the tentacles, but the body is small. This species lives at depths far below the average swimmer. The only octopus regularly fatal to humans is the small blue-ringed octopus found on the reefs of Australia.

The largest and most impressive squid is *Architeuthis princeps,* which may reach a length in excess of 18 m (60 feet). However, despite its large size it is not generally considered a threat to man as it lives in very deep water and is only known from the strandings of a few moribund specimens or from the stomachs of sperm whales which feed upon them. The Humboldt current squid lives off the coast of South America, preys upon tuna and is important to local fisheries and sports fishermen. Nevertheless, *Architeuthis* possesses large jaws which are capable of inflicting serious bites.

OCTOPUSES

Class CEPHALOPODA
Family OCTOPODIDAE

This large family consists of about 24 genera inhabiting most seas. Capable of using their coloration for camouflage and for giving signals, octopuses have extremely well-developed brains encased in cartilaginous skulls.

Octopus maculosus, *Australian blue-ringed octopus*

***Octopus lunulatus* (Quoy and Gaimard)**
Spotted octopus

DESCRIPTION Length up to about 20 cm (8 inches). Closely related to the Australian blue-ringed octopus.

HABITAT Many are found in rock pools in the intertidal zone. Generally located in the Indo-Pacific region, but probably a little farther north than *O. maculosus*.

OTHER POINTS Capable of inflicting a deadly bite.

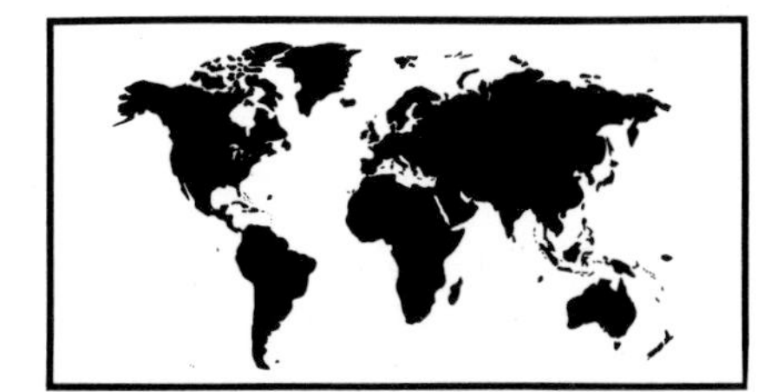

***Octopus maculosus* Hoyle**
Australian blue-ringed octopus

DESCRIPTION About 10 cm (4 inches) in diameter; small blue rings with a somewhat luminous quality on tentacles.

HABITAT Indo-Pacific area, especially common along the southern Australia coast.

OTHER POINTS Capable of inflicting a fatal bite.

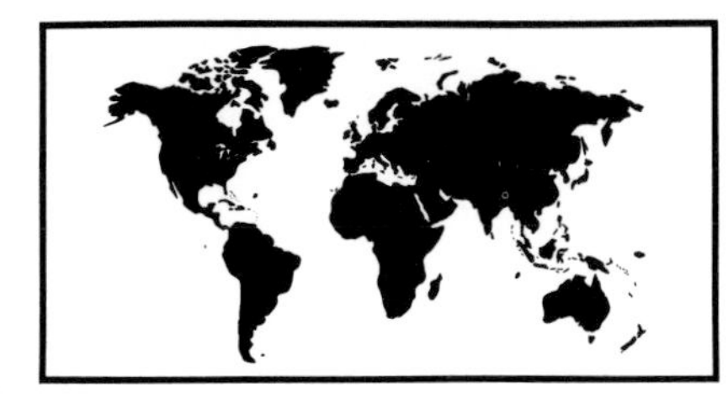

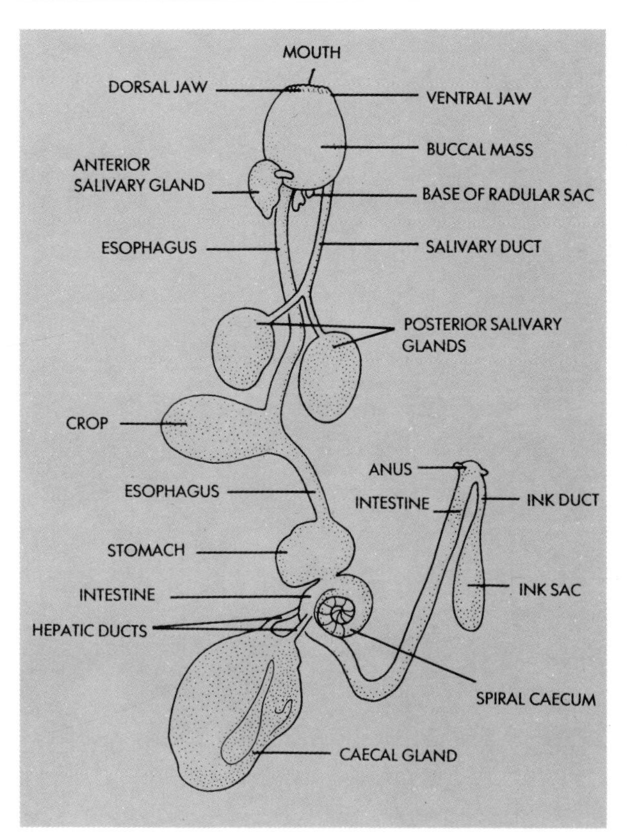

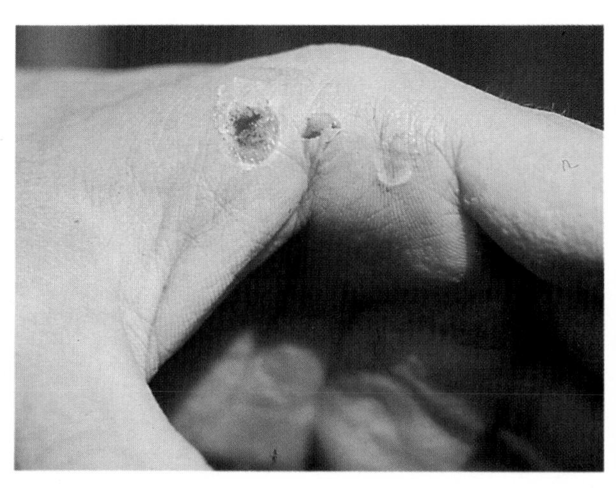

Above: Drawing of the venom apparatus of the octopus
Above right: Wound from an octopus bite

OCTOPUS BITES

Venom apparatus of the octopus

The jaws of the octopus are shaped like a parrot's beak and can bite with great force, tearing the captured food which is held by tentacles. As the octopus bites it discharges venom through its salivary glands.

Several deaths have been reported from octopus bites. Near East Point, Darwin, in Australia, a diver caught a small blue-ringed octopus, which was reported to be about 20 cm (8 inches) across. The diver let the octopus crawl over his arm and shoulder and finally to the back of his neck, where the animal stayed for a few moments. When the octopus was on his neck, it bit him and caused a small trickle of blood. A few minutes after the bite, the victim began to feel ill, and he died about two hours later.

Prevention

Underwater caves which are likely to be inhabited by octopuses should be avoided by the inexperienced diver. Wearing an outer cloth garment makes it difficult for an octopus to adhere to the skin. Regardless of their size, octopuses should be handled with gloves. Some of the smaller species seem to be the most aggressive biters. If not familiar with the species, it is best to leave well alone. If it is necessary to kill an octopus, stab it between the eyes.

Treatment

Octopus bites from most species are usually of minor concern and can be treated symptomatically, but bites from the blue-ringed or spotted octopus must be treated with urgency. First aid at the scene might include the pressure-immobilization technique described on page 60, although this is as yet unproven for management of an octopus bite. For further medical advice see Chapter 7.

MARINE BRISTLEWORMS

Phylum ANNELIDA
Class POLYCHAETA

Segmented worms, or annelids, are organisms that have a long body which is usually segmented. Each segment has two bristle-like setae, and in some species these setae can sting. Other species have tough jaws with which they inflict a painful bite. Annelid worms are usually found when turning over rocks and coral boulders.

The polychaete worms are segmented, marine bristleworms that are very abundant in most marine habitats. They are characterized by a long segmented body in which each segment has numerous bristles (or chaetae) arranged on lateral prominences of the body wall called parapodia. The bristles, which may be long and display colorful iridescence, can in some species penetrate the skin and cause a painful rash. Other worms are voracious hunters with substantial jaws on an evertable proboscis, and are capable of inflicting a painful bite. The polychaetes are less cryptic in habit and are usually found beneath rocks or burrowing in marine particulate sediments.

There are many families of polychaetes, but those most likely to prove dangerous or inconvenient to man are in the families Amphinomidae and Glyceridae. Several examples are described below.

Family AMPHINOMIDAE

This family consists of 17 genera and about 115 species. They are generally tropical or subtropical and found from shallow water to great depths. The body is short, flattened and moderately elongated.

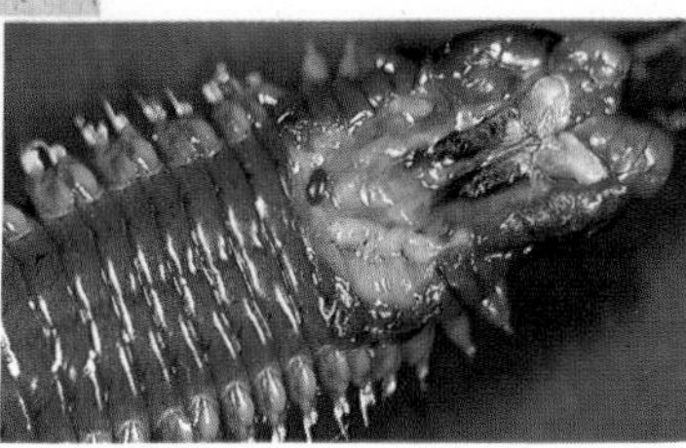

Eunice aphroditois (Pallas)
Biting reef worm

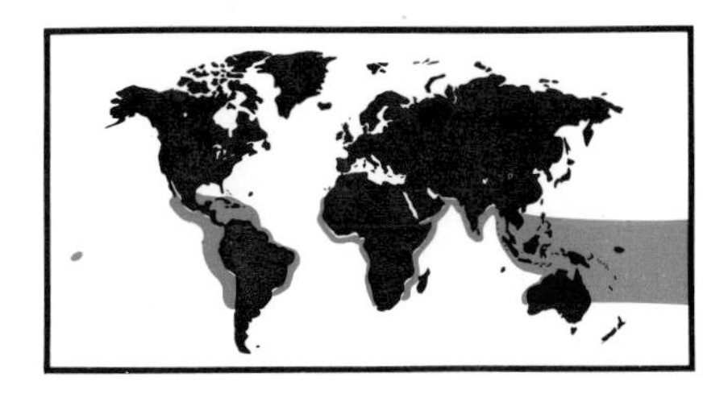

DESCRIPTION Length up to 1.5 m (5 feet); equipped with large chitinous jaws capable of inflicting a bite.

HABITAT Found in tropical areas throughout the world.

OTHER POINTS The wound may be a few millimeters in diameter, sometimes becoming swollen and inflamed. They often become infected.

***Eurythöe complanata* (Pallas)**
Bristleworm

DESCRIPTION Length 10 cm (4 inches) or more; very beautiful iridescence; short and fluffy appearance, the fluffiness comprising hundreds of sharp bristles.

HABITAT Gulf of Mexico and the tropical Pacific.

OTHER POINTS Its hollow bristles are reported to be venomous. Stings may result in intense skin swelling, with a burning sensation or numbness. The bristles can penetrate thin gloves, so these worms should be handled carefully.

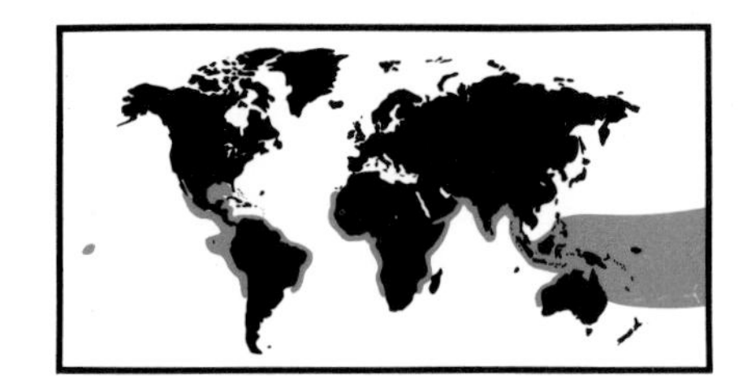

***Hermodice carunculata* (Pallas)**
Fire worm

DESCRIPTION Length up to 30 cm (12 inches); a flat, segmented body. The group of white bristles along each side are hollow and reputedly venom-filled; they easily penetrate flesh and break off if the worm is handled. The worm flares out the bristles when disturbed.

HABITAT Abundant on reefs, beneath stones in rocky or seagrass areas and on some muddy bottoms; also found on floating debris at the surface of the water. Located in Florida and the West Indies.

OTHER POINTS One of the most dangerous annelids, said to be capable of inflicting a paralyzing effect with its bristles.

Family GLYCERIDAE

This family consists of three genera and about 75 species of worldwide distribution. Species are found at varying depths; they are common in estuaries.

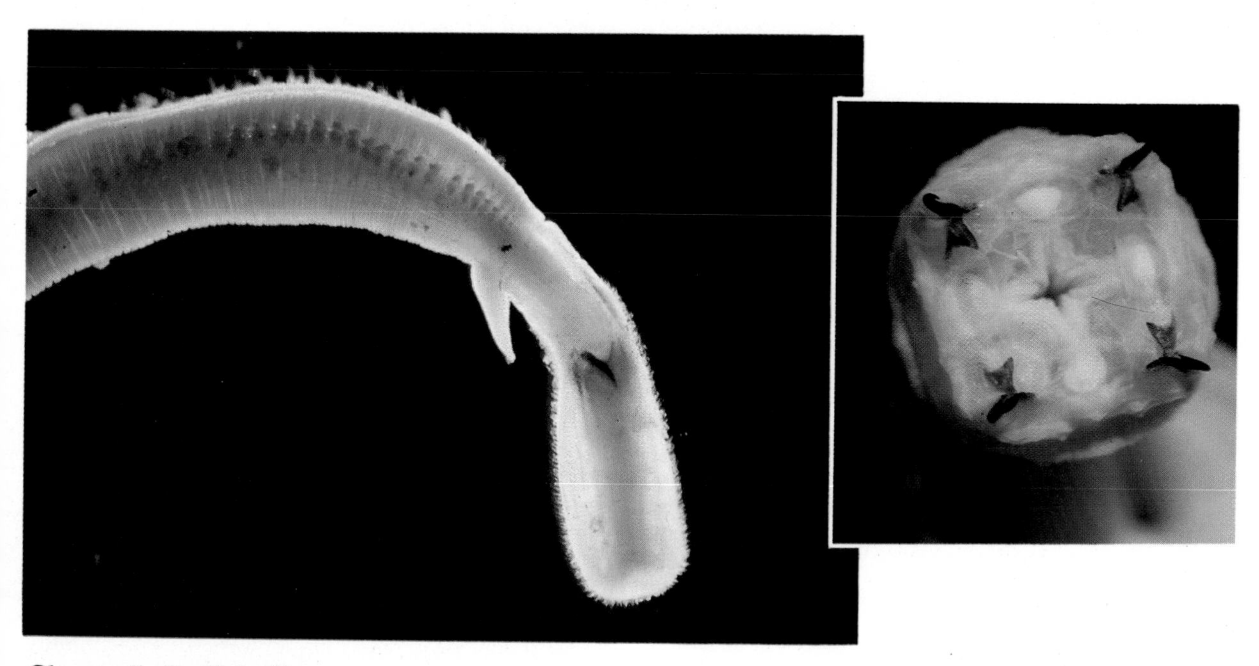

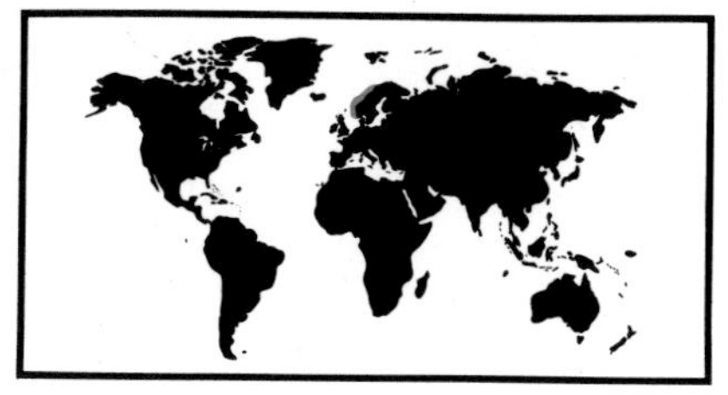

***Glycera alba* De Blainville**
Bloodworm

DESCRIPTION Body length 15 to 20 cm (6 to 8 inches). The segments are well defined and consist of two rings. The head is acutely tapered, with eight segments and four short tentacles at the tip.

HABITAT Muddy burrows in the intertidal zone and below low water areas; ranges along the coasts of the British Isles and Norway.

OTHER POINTS Both *G. alba* and *G. dibranchiata* Ehlers, a closely related species, can inflict painful bites with venomous jaws. *G. dibranchiata* ranges from North Carolina to northeast Canada, and is used in the USA and Canada as a bait worm.

BRISTLEWORM STINGS

Prevention

Divers should exercise care in handling noxious annelid worms and be wary when turning over rocks and corals. Cotton gloves will probably provide adequate protection from bloodworms, but rubber or heavy neoprene gloves are advisable when handling bristleworms.

Treatment

Treatment of bristleworm stings is mainly symptomatic. All large visible bristles should be removed with forceps. The smaller bristles can best be removed by applying adhesive tape to the skin and peeling it off. After this maneuver, acetic acid 5% (vinegar), isopropyl alcohol 40 to 70%, dilute ammonia, or a paste of unseasoned meat tenderizer (papain) may provide some pain relief. For further medical advice see the section Marine Worm Trauma in Chapter 7.

ECHINODERMS

Phylum ECHINODERMATA
Class ASTEROIDEA

The echinoderms are a very large group of marine invertebrates, characterized by radial symmetry as adults, often with a pentameous (five-rayed) body form. The body is supported on an internal calcareous skeleton of ossicles, spines, or plates. Movement is carried out by means of tube feet located on each arm or ray. These are controlled by water vesicles on the inside of the body, which when contracted cause the numerous tube feet to extend. The feet are used to open bivalve mollusks upon which most starfish feed, or as a means of transferring food to the centrally placed mouth. While sea urchins browse the substrate using calcareous jaws, the starfish consumes its prey by everting its stomach around the organism and digesting it. Only the classes Asteroidea (starfishes) and Echinoidea (sea urchins) are likely to be serious hazards to swimmers.

STARFISHES

Family ACANTHASTERIDAE

Members of this family have 12 to 18 arms or rays, the number increasing with size. Large pointed spines are present on the margins and upper surface.

Acanthaster planci **(Linnaeus)**
Crown of thorns starfish

DESCRIPTION May be more than 60 cm (24 inches) in diameter; has 13 to 16 arms. The top of the starfish is covered with many long, sharp and strong spines about 6 cm (2.3 inches) or more in length. The spines have a special thick integumentary sheath which holds dangerous venom. These may cause painful wounds and may break off in the wound.

HABITAT Lives on coral reefs in the Indo-Pacific, from Polynesia to the Red Sea; where abundant, may destroy entire coral colonies.

OTHER POINTS The only known venomous asteroid. A closely related species, *A. ellisi*, is found in the Gulf of California.

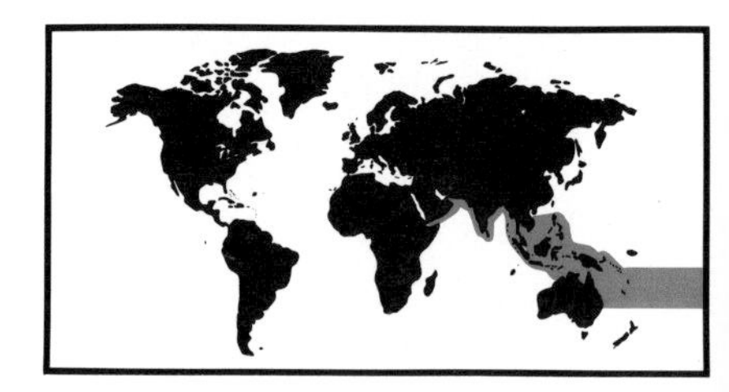

STARFISH STINGS

Prevention

The upper surface of the starfish is covered with numerous venomous spines. Care should be taken when handling these animals, and heavy gloves should be worn. Contact with bare hands is only safe on the animal's soft underside.

Treatment

Exposure of the wound to the suction pads of a living starfish is said to be helpful in alleviating the pain; we can neither verify nor condemn this remedy. The wounds should be immersed in hot water to tolerance (45°C or 113°F), taking care not to scald the victim, for 30 to 90 minutes or until there is significant pain relief. All easily extracted spines should be removed. For further medical advice see Chapter 7.

SEA URCHINS

Class ECHINOIDEA

Sea urchins, or "sea eggs", are common in all seas and are found at all depths. The body is globe-shaped, although it may carry depressions. Covered in sharp spines, they can be hazardous to swimmers in shallow reef areas. They live mostly in crevices, but may emerge at night to feed and browse on surrounding rocks. Movement is by means of five rows of tube feet. Some species burrow in the sand.

Family DIADEMATIDAE

This family contains 29 species in eight genera, distributed worldwide in tropical and subtropical seas, mostly in the western Indo-Pacific.

***Diadema antillarum* Philippi**
Long-spined, hairy sea urchin

DESCRIPTION Diameter of body about 10 cm (4 inches); body is black and spines are black or white. The spines are relatively long, sometimes reaching 40 cm (16 inches) with slender points that easily penetrate human skin.

HABITAT During the day hides within crevices or around sheltered locations on the reef; at night it moves into the open, often onto sandy areas. May also be abundant in seagrass beds, rocky shelves, mangrove areas and sandy bottoms. Found in the West Indies.

OTHER POINTS The needle-like spines are hollow and may contain a poison. Injuries from spines are extremely painful. Because of their friable nature, the spines are difficult to remove.

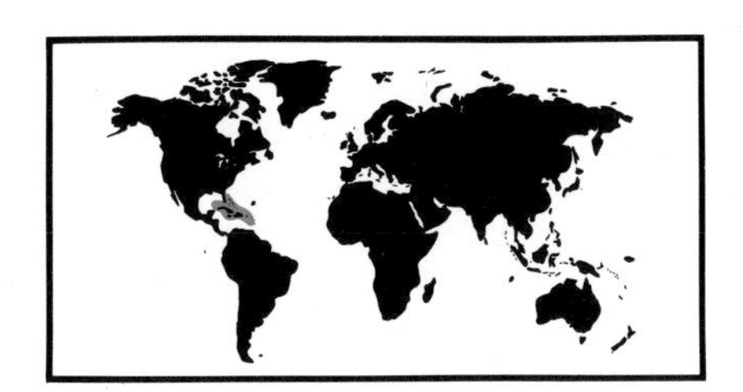

***Diadema setosum* (Leske)**
Long-spined or black sea urchin

DESCRIPTION Similar to *D. antillarum*, except for the characteristic features of this Pacific form, which are isolated blue spots that are visible to a greater or lesser extent. Spines are generally black, reddish or greenish.

HABITAT Indo-Pacific area, from Polynesia to east Africa.

OTHER POINTS Dangerous species to handle.

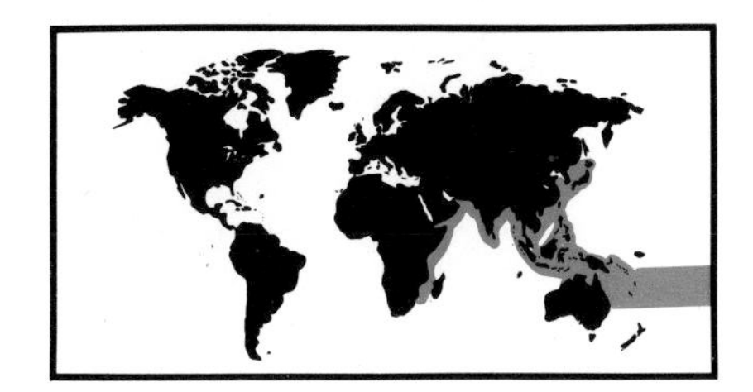

Family ARBACIIDAE

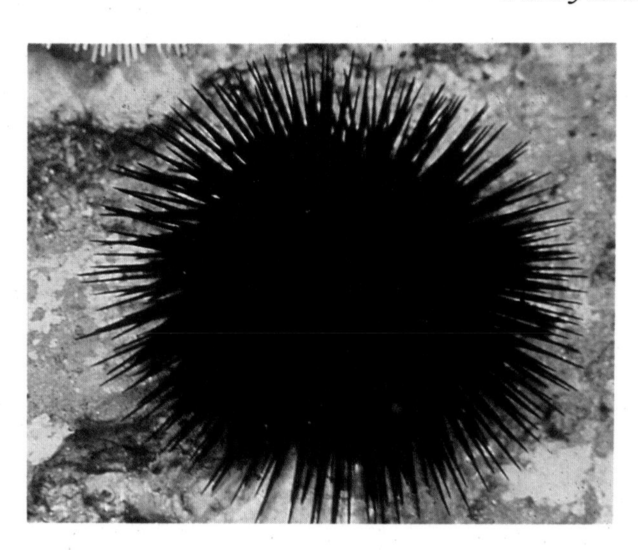

***Arbacia lixula* (Linnaeus)**
Sea urchin

DESCRIPTION Diameter up to 5.8 cm (2.2 inches), height 2.5 cm (1 inch) and spines of equal length; color dark purple.

HABITAT Hard substrates along the surf in the Mediterranean Sea, the west coast of Africa, the Canaries, Madeira, the Azores and Brazil. Nocturnal, feeding on calcareous algae.

OTHER POINTS Venom is toxic to a variety of invertebrate and vertebrate animals. Effects of the venom on humans are unknown.

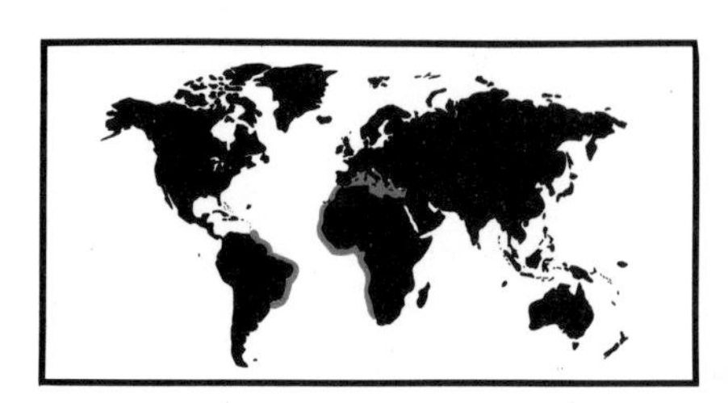

Family ECHINOTHURIDAE

***Asthenosoma varium* Grube**
Leather urchin

DESCRIPTION Up to 17 cm (6.7 inches) in diameter; alternating purple, brown and black coloration. Venomous pedicellariae.

HABITAT Deep water or tidal zones throughout the Indo-Malayan archipelago, Indian Ocean and the Gulf of Suez.

OTHER POINTS Its spinal venom organs appear to be developed in the secondary aboral spines. They are composed of connective and muscular fibers which enclose a venom sac surrounding the tip of the spine.

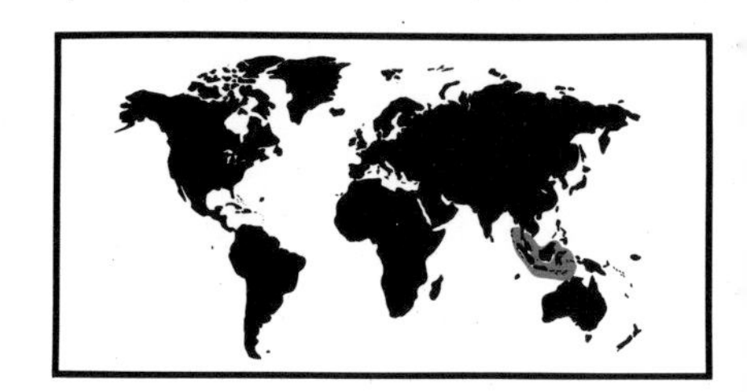

Family STRONGYLOCENTROTIDAE

Strongylocentrotus franciscanus **(Agassiz)**
Red sea urchin

DESCRIPTION Much larger than others of the genus, reaching 15 cm (6 inches) or more in diameter; usually red or red brown, sometimes bright purple. Somewhat uncommon.

HABITAT Very low intertidal zones, on open, coastal rocky shores; more abundant subtidally, and extending to depths of 90 m (300 feet). Ranges from northern Japan and Alaska to Baja California.

OTHER POINTS Large adults may be 20 years old. Spines are strong and can readily penetrate skin.

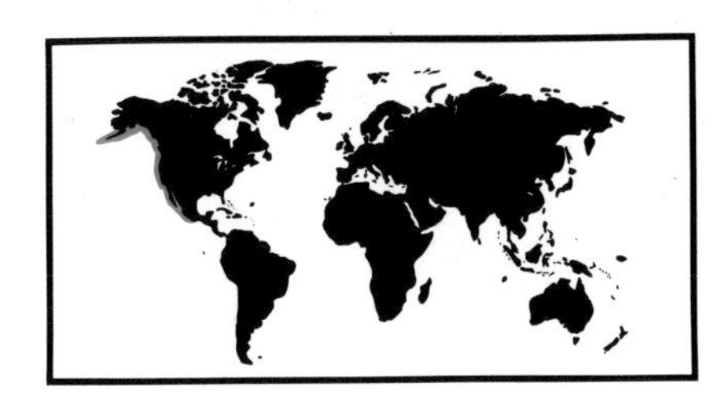

Family TOXOPNEUSTIDAE

The family contains 35 species in 10 genera. Most species attain sizes of 10 to 15 cm (4 to 6 inches) in body diameter and are a conspicuous part of shallow-water epifauna where they occur.

Toxopneustes pileolus **(Lamarck)**
Felt cap sea urchin

DESCRIPTION Body diameter 13 cm (5 inches), with relatively short spines. When the pedicellariae are extended, the species may appear as a brilliantly-colored dense bed of "small flowers", varying in color combination between purple, yellow, red, green or white. Hidden beneath are the venom-clad spiny jaws of the globiferous pedicellariae.

HABITAT Throughout the Indo-Pacific region, from Malaysia to east Africa and Japan.

OTHER POINTS Its poison has a direct action on the nervous system.

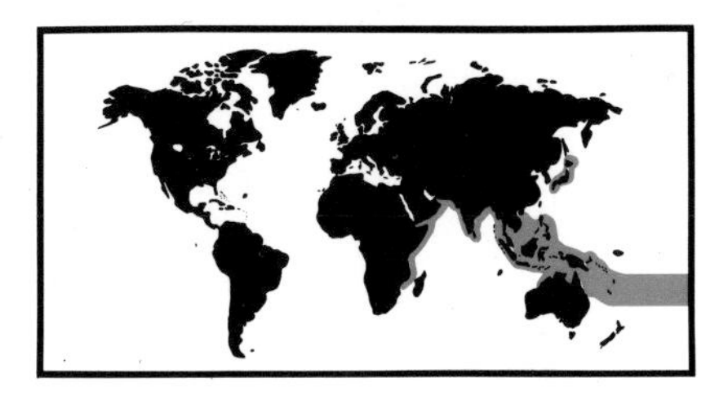

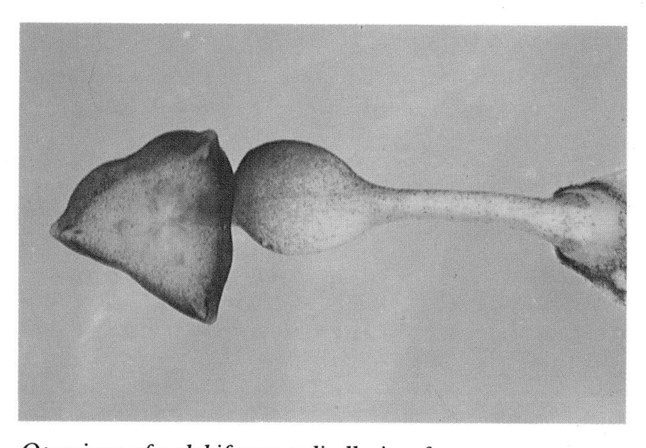

Open jaws of a globiferous pedicellarium from ***Sphaerechinus granularis***

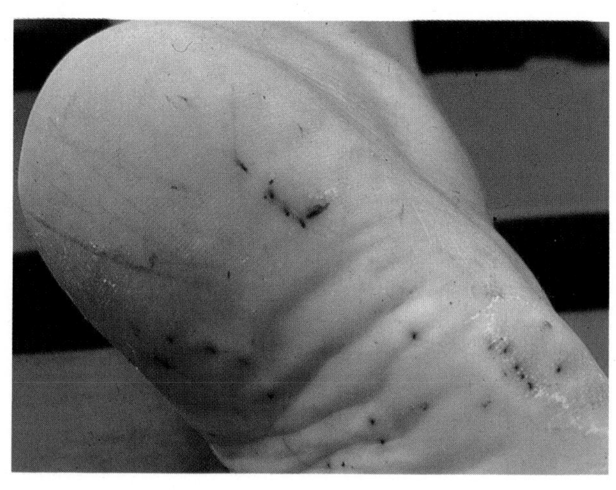

Wound from the spines of a sea urchin

SEA URCHIN STINGS

Venom apparatus of sea urchins

There are two types of venom organs in sea urchins: venomous spines and pedicellariae. Most of the dangerous species have one or the other, but seldom both.

The spines of sea urchins vary greatly from group to group. In most cases they are solid, have blunt and rounded tips, and do not have a venom organ. However, some species have long, slender, hollow, and sharp spines which are very dangerous to handle. Because of extreme brittleness, the spines may break off in the wound and become difficult to remove. The spines of the black sea urchin, *Diadema setosum*, may reach a length of 30 cm (12 inches).

Pedicellariae are small and delicate seizing organs, which are found among the spines upon the shell. There are several different types. One of these, because of its globe-shaped head, is called the globiferous pedicellarium and serves as a venom organ. The pedicellarium is made up of two parts: a head and a supporting stalk. One of the main functions of pedicellariae is defense. The pedicellariae do not release their hold as long as the object moves, and if it is too strong to be held, they are torn from the shell. Pedicellariae that have been torn from the shell may remain alive for several hours and continue to envenomate the victim.

Prevention

A sea urchin with elongated, needle-like spines should not be handled. Leather and canvas gloves, shoes, and flippers do not guarantee protection. Care should be taken in handling any tropical species of short-spined sea urchin when not wearing gloves. It is also necessary to be careful when lifting rocks or poking into crevices, and moving among corals where sea urchins abound.

Treatment

The treatment of sea urchin spine wounds varies with the type of spine and the body region involved. Immersion into hot water may provide some relief from pain. The wound should be immediately immersed in hot water to tolerance (45°C or 113°F) for 30 to 90 minutes or until there is significant pain relief, taking care to avoid scalding the victim. Any detached pedicellariae still attached to the skin must be removed mechanically or envenomation will continue. This may be accomplished by the application of shaving foam and gentle scraping with a razor. Embedded spines should be removed carefully, as they are easily fractured. Many sea urchin spines may break off when grasped with a pair of forceps and often it is difficult to remove the spines without surgical intervention. For further medical advice see Chapter 7.

CHAPTER 3

VENOMOUS VERTEBRATES

Fish are the most diverse group of vertebrates in the world, ranging in habitat from high mountains and hot thermal springs to the deepest ocean depths. With over 27,000 species, it is not surprising that fish have evolved numerous defensive mechanisms that are potentially dangerous to man. Spines, which are often accompanied by venom, are found in many groups.

In some regions venomous fish can pose a serious threat to swimmers, divers and fishermen. Much of the risk is diminished if the habitats and habits of the species are known. This chapter identifies and describes a number of key species and representative groups.

Another group of venomous marine vertebrates is the sea snakes, which are widely distributed in the Indo-Pacific. Sea snakes are beautifully adapted for life at sea, with flattened tails for locomotion. The body is often conspicuously colored. They belong to a group of snakes in which the fangs are located at the front of the mouth and are small and easily dislodged.

SPINY DOGFISH

Phylum CHORDATA
Class ELASMOBRANCHII (Chondrichthys)
Family SQUALIDAE

Squalidae consists of a number of genera, but only the spiny dogfish or spur dog is thought to inject poison into wounds. Most dogfish are slow-moving and may spend long periods resting on the bottom, but members of Squalidae are mostly deepwater active species, some of which undergo extensive migrations. Spiny dogfish are viviparous.

Squalus acanthias **Linnaeus**
Spiny dogfish, spurdog

DESCRIPTION Length up to 1.5 m (5 feet); body typically shark-shaped, fairly slender with a pointed snout. Gray coloring above with white below; may have white markings on the sides. The first dorsal fin is larger than the second, and in front of each is a spine.

HABITAT Found on both sides of the North Atlantic and North Pacific Oceans. Can occur in large schools, feeding on commercially important mackerel and herring species. Close relatives are found throughout temperate and tropical seas.

VENOM APPARATUS OF SPINY DOGFISH

Wounds from the spiny dogfish are caused by the dorsal spine, which is located at the front part of each of the dorsal fins. The venom gland appears as a shiny, whitish structure located on the upper portion of each spine. Venom is introduced into a wound when the spine punctures the skin. Stings usually occur from handling of dogfish, so commercial fishermen need to be particularly careful. For medical advice see Treatment of venomous fish stings, page 118.

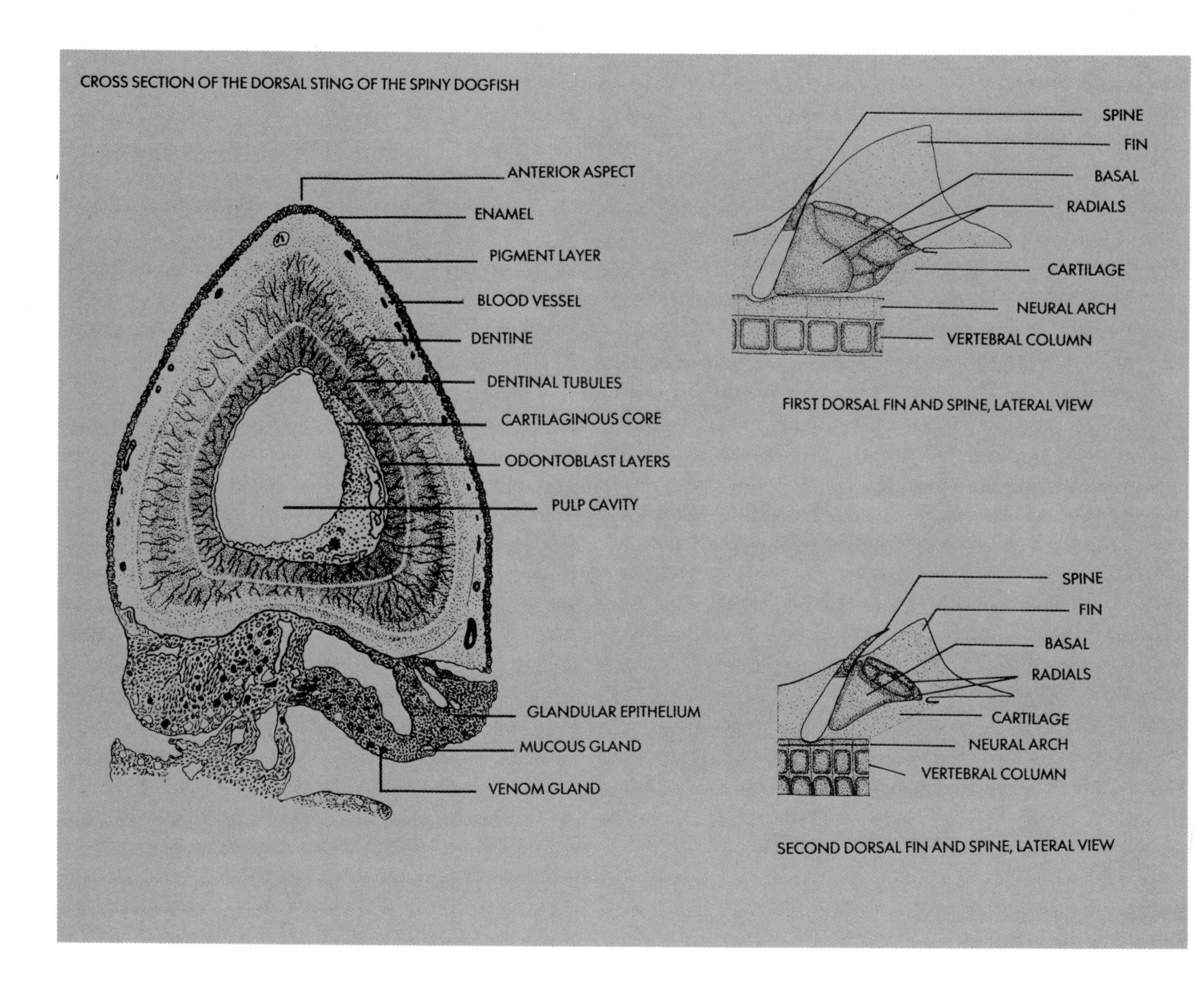

STINGRAYS

Marine stingrays are distributed within six families: Dasyatidae (stingrays or whip rays); Gymnuridae (butterfly rays); Mobulidae (devil rays or mantas); Myliobatidae (eagle rays or bat rays); Rhinopteridae (cownosed rays); and Urolophidae (round stingrays).

Rays are commonly found in tropical, subtropical and warm temperate seas. With the exception of the South American river rays (Potamotrygonidae), which live in fresh water, rays are mainly marine animals. Although most common in shallow water, rays have been known to swim in moderate depths. A deep sea species has been reported in the central Pacific Ocean. Sheltered bays, shallow lagoons, river mouths and sandy areas between patch reefs are favorite places. They may be seen lying on top of the sand, or partially covered by the sand, with only the eyes, breathing holes and a portion of the tail showing above the sand. Rays burrow into the sand and mud, feeding on worms, mollusks and crustaceans.

Most stingrays are capable of inflicting severe wounds if accidentally disturbed by a person wading or swimming. Since in tropical seas they are commonly found in shallow inshore waters, there are many records of stingray attacks, with about 1500 each year from the USA alone.

Family DASYATIDAE

This is a very large family of rays, comprising five genera with 90 to 100 species. The tail is long and slender, with a serrated spine which can inflict a serious wound. The spine is grooved and also venom-bearing.

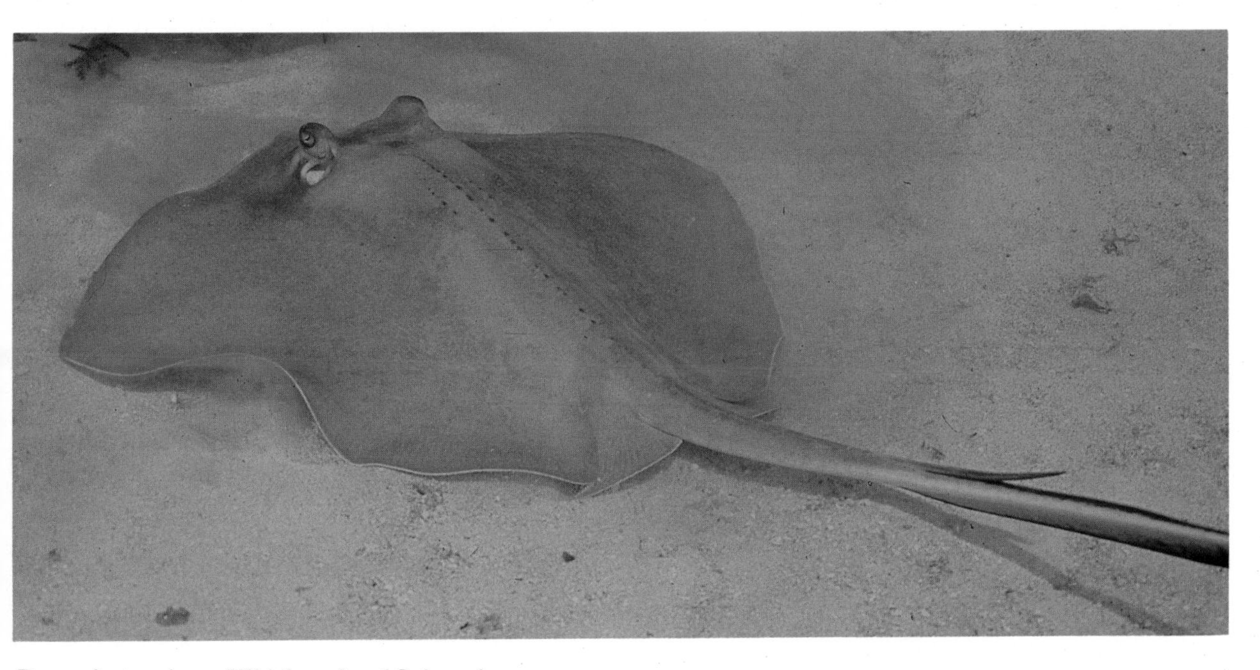

Dasyatis americana **Hildebrand and Schroeder**
Southern stingray

DESCRIPTION Length 2 m (6.5 feet).

HABITAT A common West Indian species that inhabits the western Atlantic from New Jersey to southern Brazil.

OTHER POINTS Equipped with a well-developed serrated spine and capable of inflicting a painful laceration.

Dasyatis brevicaudata **(Hutton)**
Giant stingray of Australia

DESCRIPTION Reported to be the largest stingray in the world, reaching a length of 4.5 m (15 feet) and a width of 2.2 m (7 feet), weighing more than 324 kg (714 pounds). Usually dark brown or gray-brown but takes on a darker color when removed from the water; some have signs of mottling or marbling.

HABITAT Tends to inhabit the bottoms of bays and estuary flats, shoal lagoons, river mouths or patches of sand between coral heads. Sometimes difficult to detect because it lies partially buried in mud or sand. Usually found in the Indo-Pacific area.

OTHER POINTS Dangerous. The venomous spine from one specimen measured 37 cm (14.5 inches). May cause fatalities.

Dasyatis kuhli **(Müller and Henle)**
Blue-spotted stingray

DESCRIPTION About 25 cm (10 inches) in width with a comparatively short tail. Beautiful appearance, covered on the upper surface of its body with distinctive, large blue spots on a golden brown or gray background.

HABITAT Usually found partially buried in sand in shallow coral reef areas. Wide geographical range in the Indo-Pacific, India and Japan.

OTHER POINTS The venomous spine can inflict a painful wound.

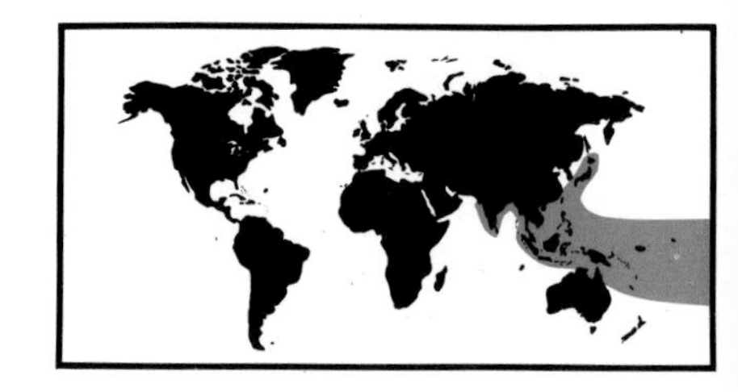

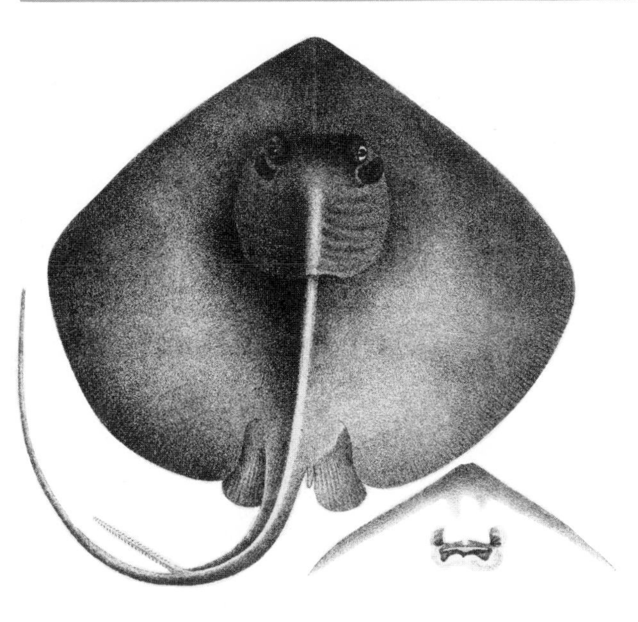

Dasyatis pastinaca **(Linnaeus)**
European stingray

DESCRIPTION Length up to 2.5 m (8.2 feet) and width of 1.4 m (4.5 feet); the back is brown and the underside white. The tail is slender and whiplike and possesses a stinging spine near the top.

HABITAT Common coastal European stingray which is also found in lagoons and brackish water, but not below a depth of 60 m (200 feet). Inhabits the northeastern Atlantic Ocean, Mediterranean Sea and Indian Ocean.

OTHER POINTS Mentioned by Pliny, who wrote that the "pastinaca marina" was able to kill a tree with its powerful sting.

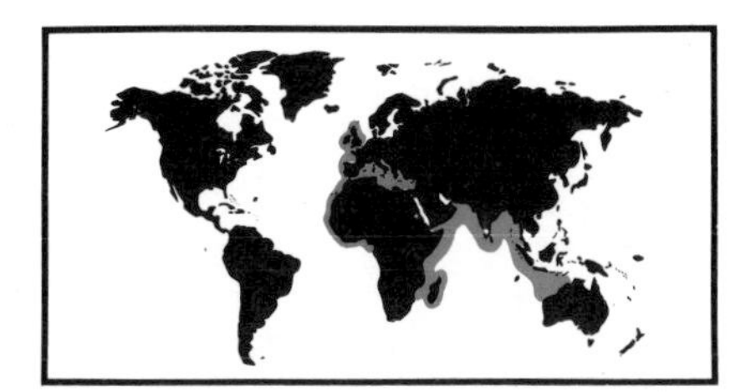

Taeniura lymma **(Forskål)**
Stingray

DESCRIPTION Length of more than 2 m (6.5 feet); the back is medium to light-brown in coloring with large light blue spots, and with a light blue band along each side of the tail.

HABITAT Shallow water in the Indo-Pacific, Australia, Indian Ocean and Red Sea.

OTHER POINTS Often kept in aquariums.

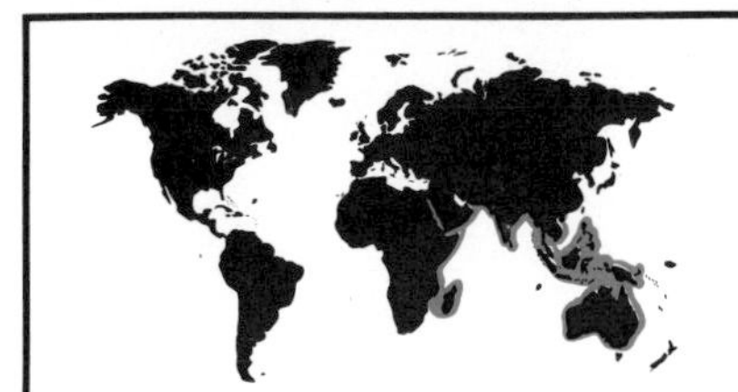

Family GYMNURIDAE

This family consists of two genera (*Aetoplatea* and *Gymnura*), comprising 12 species. These rays are found worldwide in tropical and warm temperate latitudes, along the shoreline to a depth of some 55 m (180 feet).

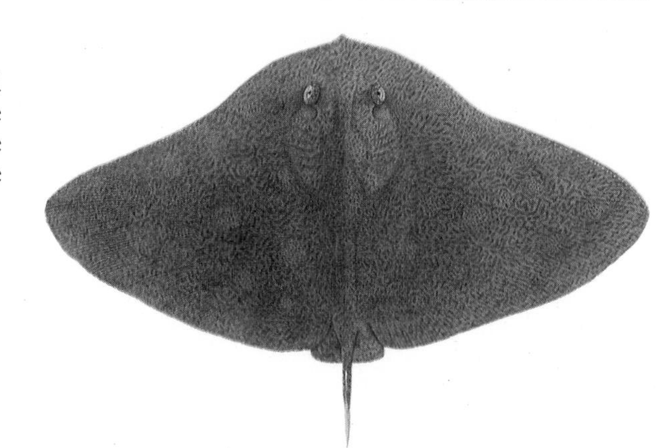

Gymnura marmorata **(Cooper)**
California butterfly ray

DESCRIPTION Wider than it is long, with a width of up to 2.2 m (7.2 feet); the gray, brown, purple, or green markings which lace its back make it unusually colorful and graceful in appearance.

HABITAT Usually found on the sea bed, from Point Conception, California, south to Mazatlan, Mexico.

OTHER POINTS Has a small venom apparatus which has limited use as a defense organ. Few records of persons being seriously hurt by this species.

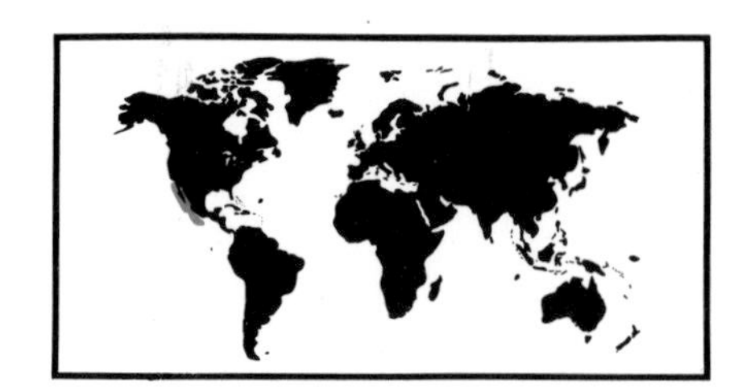

Family MYLIOBATIDAE

Members of the family Myliobatidae, the eagle rays, are characterized by their pectorals, which are either narrow and opposite the eyes or entirely interrupted at that point, the head thus being conspicuously marked off from the rest of the body. The family contains approximately 22 species in four genera (*Aetobatus, Aetomylaeus, Myliobatis* and *Pteromylaeus*).

Aetobatus narinari **(Euphrasen)**
Spotted eagle ray

DESCRIPTION Weighs up to 225 kg (500 pounds) and attains 2.3 m (7.5 feet) in width; numerous white or bluish spots scattered over the upper surface of the body. Can have as many as five stings at the base of its long tail.

HABITAT Spend much of their time swimming over the sea bed or near the surface of the water with a flying-like motion. An inshore species ranging throughout tropical and warm temperate regions of the Atlantic, Indo-Pacific, and Red Sea.

OTHER POINTS An example of a multi-spined stingray. Large groups swimming in formation can be an awe-inspiring sight.

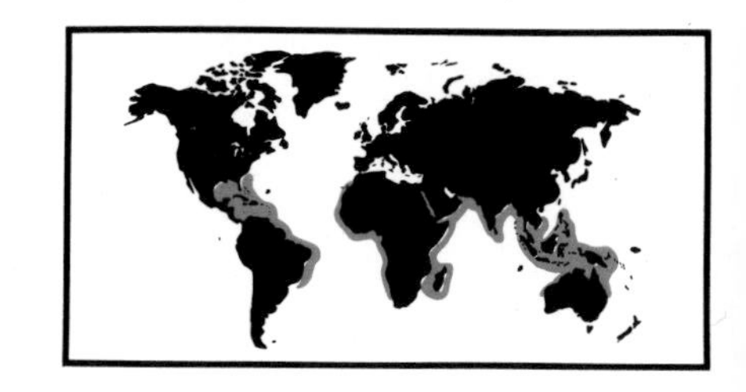

***Myliobatis californicus* Gill**
California bat stingray

DESCRIPTION Width up to 1.2 m (4 feet) and weight over 68 kg (150 pounds); dark brown to dark olive or almost black above, white below.

HABITAT Fairly common; can be observed in open areas near kelp forests. Ranges from Oregon to Baja California.

OTHER POINTS The teeth are particularly suited to crushing oysters, clams and crustacea. It is very destructive to oyster beds.

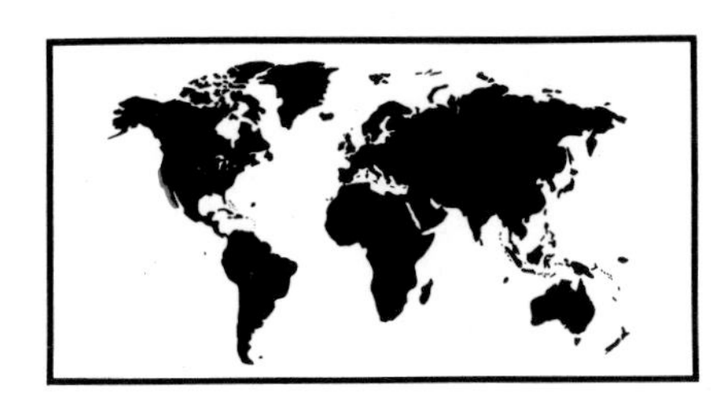

Family RHINOPTERIDAE

This family consists of a single genus with 10 species which occur worldwide, except around the islands of the western Pacific. The cownose rays differ from the eagle rays in possessing a pair of subrostal lobes or fins. Some authors include them in a single family due to their similar morphological characteristics.

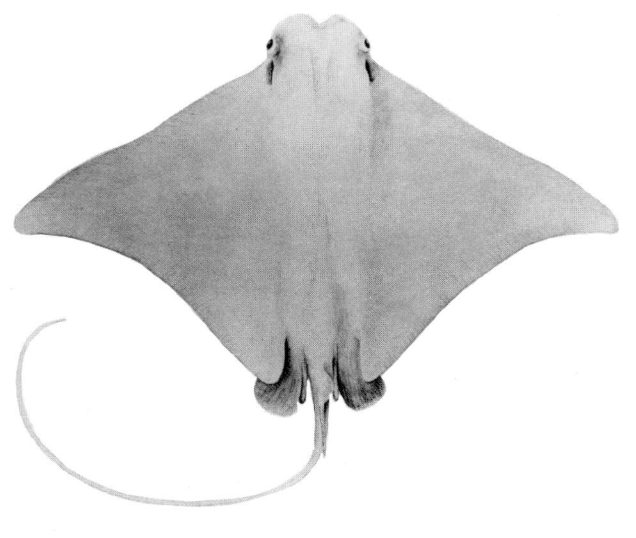

***Rhinoptera bonasus* (Mitchill)**
Cownose ray

DESCRIPTION Width up to 2 m (6.5 feet) and weight up to 45 kg (100 pounds).

HABITAT Frequently observed swimming in small schools just above the ocean floor. Found along the coastal western Atlantic, from New England to Brazil.

OTHER POINTS It has either one or two dangerous spines near the base of its long, thin tail.

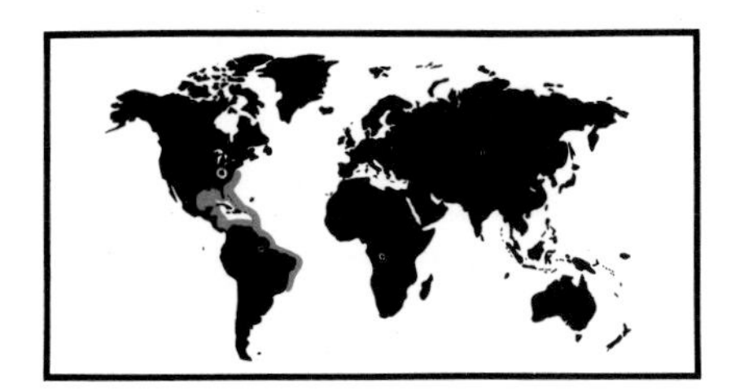

Family UROLOPHIDAE

This family is made up of two genera containing 30 species. The round stingrays have a reputation for their ability to strike an object with some precision using their powerful muscular tails. This family is responsible for most cases of human envenomation by stingrays.

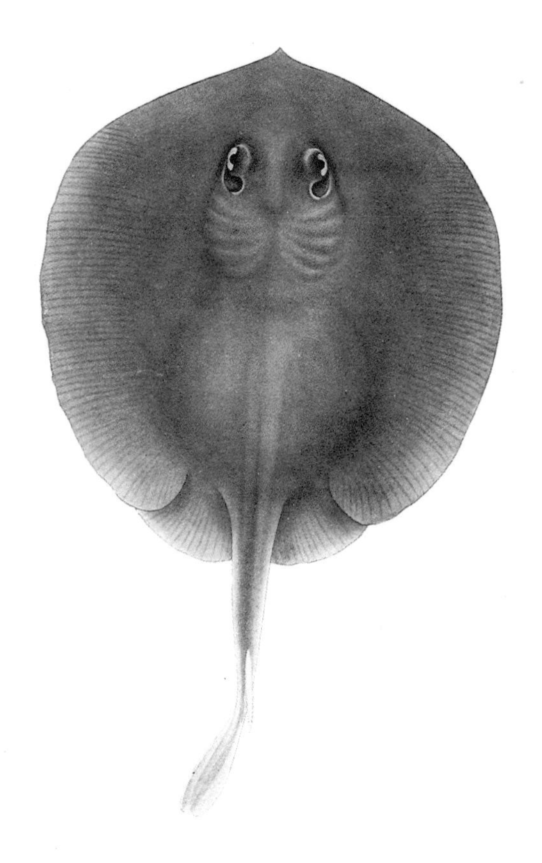

***Urolophus halleri* Cooper**
Round stingray

DESCRIPTION Length 50 cm (20 inches); shades of brown on the dorsal surface; sometimes mottled, spotted and yellowish below. The disk shape of the body is nearly circular. There is no dorsal fin.

HABITAT The most common California stingray, located from Point Conception, California, to the Panama Bay. Usually found buried in the mud.

OTHER POINTS Despite their comparatively small size, they can inflict painful wounds.

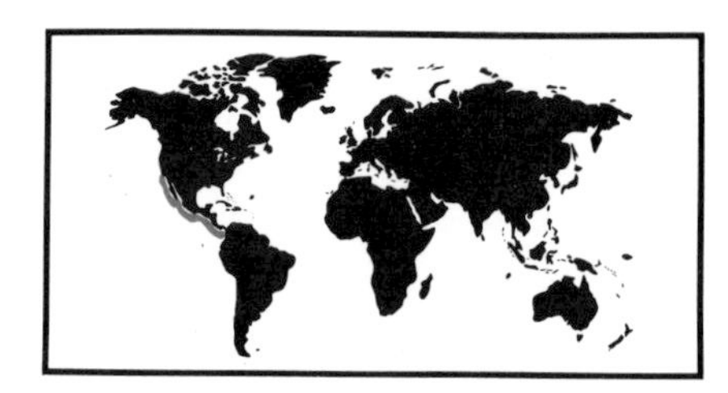

***Urolophus jamaicensis* (Cuvier)**
Yellow stingray

DESCRIPTION Length up to 67 cm (26 inches). Similar in appearance to *U. halleri*.

HABITAT Shallow waters with muddy or sandy bottoms in the western tropical Atlantic, from Florida southward to the Caribbean.

OTHER POINTS Commonly found in Jamaican waters. Local fishermen pay particular respect to it, as it is known to be capable of inflicting dangerous wounds with its venomous spine.

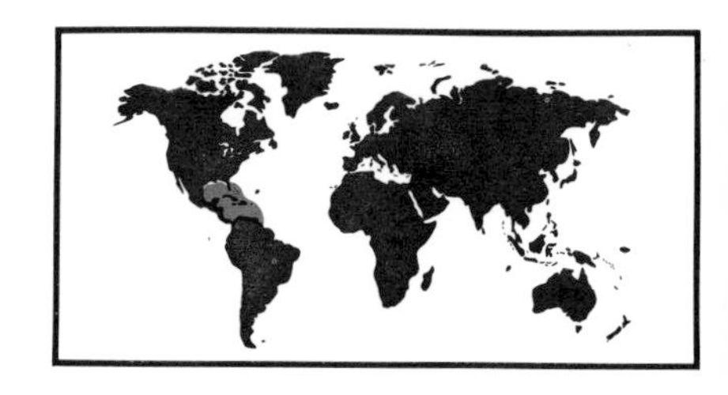

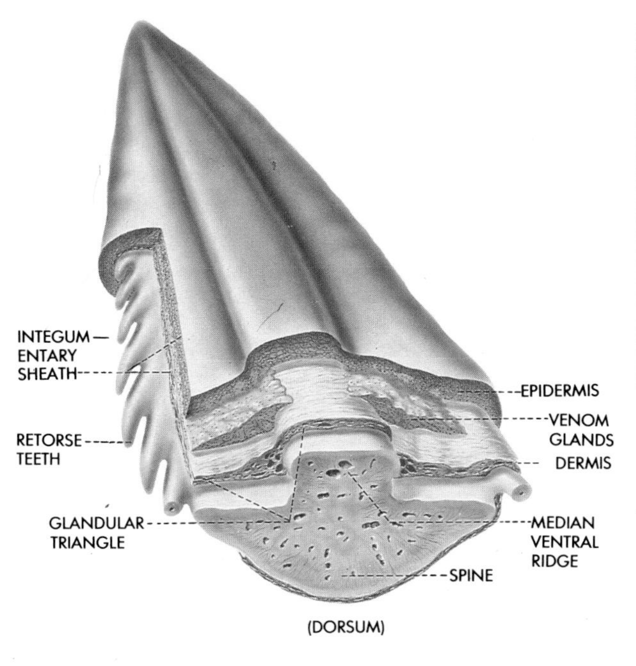

Gross anatomy of a typical stingray venom apparatus

***Dasyatis americana** buried in sand*

STINGRAY ATTACKS

Venom apparatus of stingrays

A study of stingrays shows there are four general types of venom organs, which vary in effect. The Gymnurid type is found in the butterfly rays. The sting is small, poorly developed and situated close to the base of the short tail, making it a feeble striking organ when compared with other types. The Myliobatid type is found in the bat and eagle rays, which have long, whiplike tails. The spines in these rays are frequently large and well developed, but situated near the base of the tail. The Dasyatid type is found in the stingrays and river rays. The spine is well developed and located away from the base of the tail, making it potentially more dangerous as a striking organ. Stingrays having this type of venom apparatus are among the most dangerous. Finally the Urolophid type is found among the round stingrays. The tail to which the spine is attached is short, muscular, and well developed. Round stingrays can inflict a serious injury with a lash of the tail.

In general, the venom apparatus of stingrays is made up of a serrated spine contained within a thin layer of tissue called the integumentary sheath. Stingrays normally have only one spine, but sometimes more than one spine is present at the same time. The spine can become detached from the stingray when it is used to sting, through wear or due to an injury. There is no evidence to support the idea that the spines are shed each year.

The spine is composed of hard, bone-like material called vasodentine. Along both sides of the spine is a series of sharp, recurved teeth. A number of irregular, shallow furrows run almost the length of the spine. Along either edge on the underside of the spine are deep grooves. These grooves contain a strip of soft, spongy and grayish tissue which produces most of the venom.

Prevention

There are about 1800 stingray attacks reported in the USA each year. It should be kept in mind that stingrays commonly lie almost completely buried in the upper layer of a sandy or muddy bottom. They are, therefore, a hazard to persons wading in water inhabited by them, the chief danger being stepping on one that is buried. Pushing or shuffling one's feet along the sandy bottom reduces the danger of stepping on a stingray as it is likely to be frightened away when approached. It is recommended that a stick be used to probe in front when wading in shallow inshore waters.

Treatment

No specific antivenin is available. See Treatment of venomous fish stings, page 118.

RATFISHES OR ELEPHANTFISHES

Ratfishes, elephantfishes or chimaeras are the various names for the same group of cartilagenous fishes having a single outside gill opening on either side of the body, which is covered by a skin fold which leads to the gill chamber. These fishes have a more or less compressed body which tapers into a slender tail, the snout being rounded or cone-shaped, and extended as a long, pointed beak. This long beak-like extension can also have a curious hoe-shaped, somewhat longer and flexible snout.

Family CHIMAERIDAE

The family Chimaeridae consists of two genera (*Chimaera* and *Hydrolagus*) containing about 21 species. Chimaeras have a wide geographical range, extending from northern to southern temperate waters. However, they seem to prefer cooler temperatures and live mostly in deeper water.

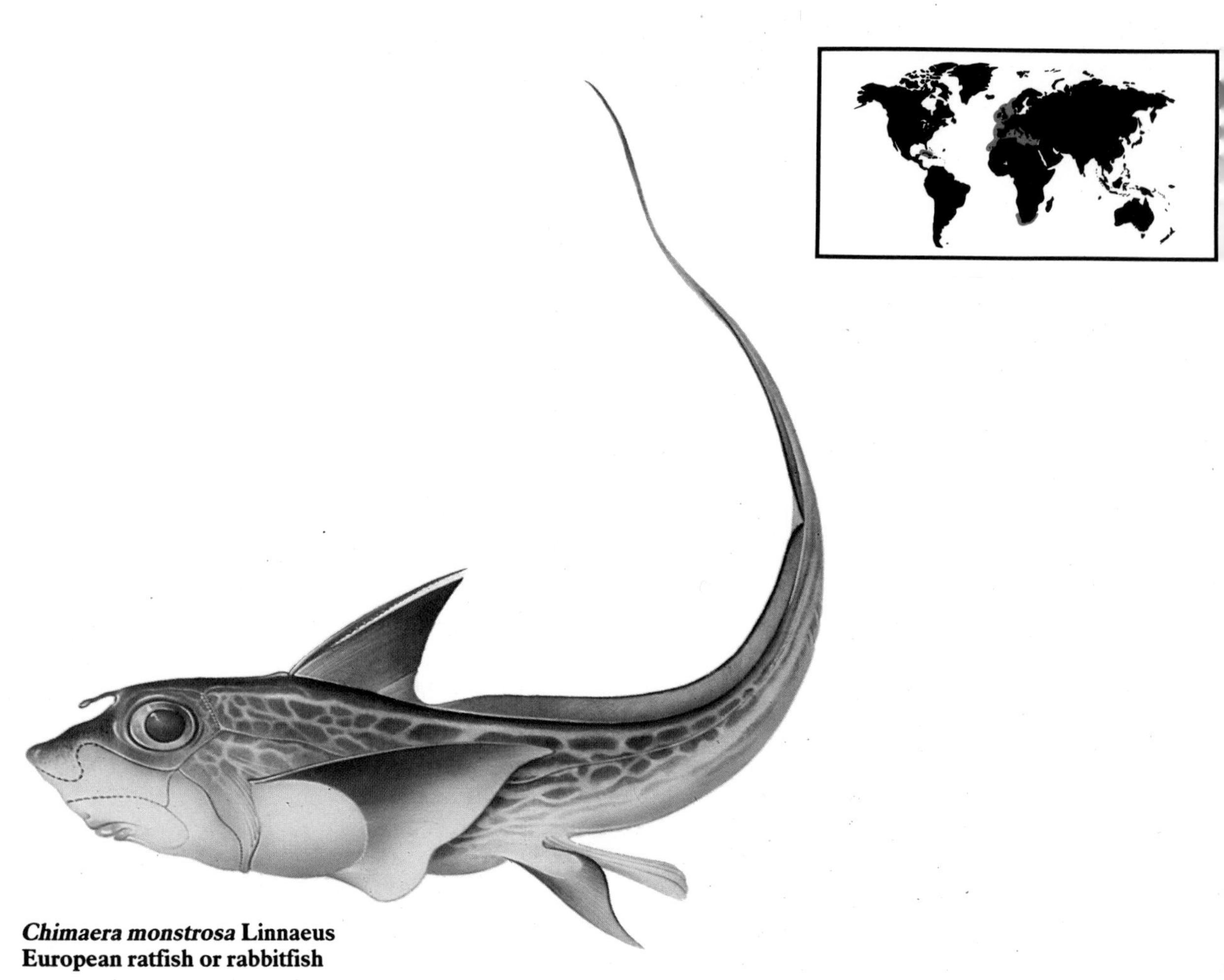

***Chimaera monstrosa* Linnaeus**
European ratfish or rabbitfish

DESCRIPTION Length 1.5 m (5 feet); silver-gray with shades of violet on the upper side with a dark marble pattern. The beak-like teeth plates of the jaw are used to mash hard-shelled prey.

HABITAT A bottom-dweller generally found on the continental shelf to depths of 200 m (650 feet), but sometimes more. Distributed in the North Atlantic from Norway and Iceland to Cuba, the Azores, Morocco, the Mediterranean Sea and South Africa.

OTHER POINTS The only *Chimaera* to inhabit European waters. Its stinging spine can be dangerous. The single dorsal spine is sharp and pointed, and although only mildly venomous can inflict a painful wound.

Hydrolagus colliei **(Lay and Bennett)**
Pacific ratfish

DESCRIPTION Length up to 1 m (39 inches); silver color with iridescent reflections of gold, blue and green pale spots on back.

HABITAT Moderately shallow water in the northern latitudes, but lives deeper in the south. Ranges along the Pacific coast of North America.

OTHER POINTS The spine can be dangerous and cause a painful wound. Fishermen are reputed to fear the jaws of the ratfish more than they do the dorsal spine.

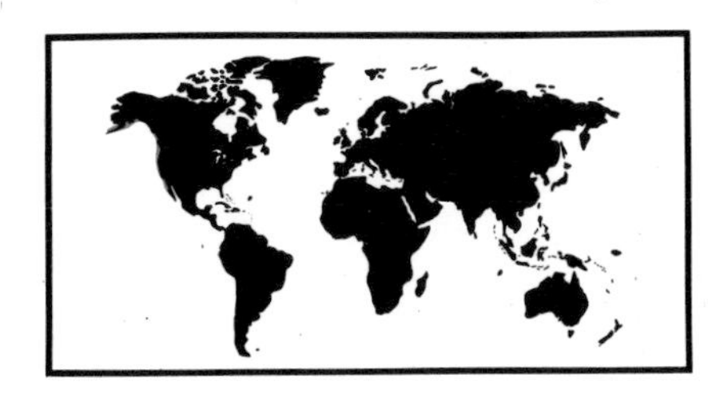

VENOM APPARATUS OF RATFISHES

The venom apparatus of ratfishes is made up of the single dorsal sting which is situated along the front of the first dorsal fin. Along the back of the spine is a shallow depression which contains the venom gland. For medical advice see Treatment of venomous fish stings, page 118.

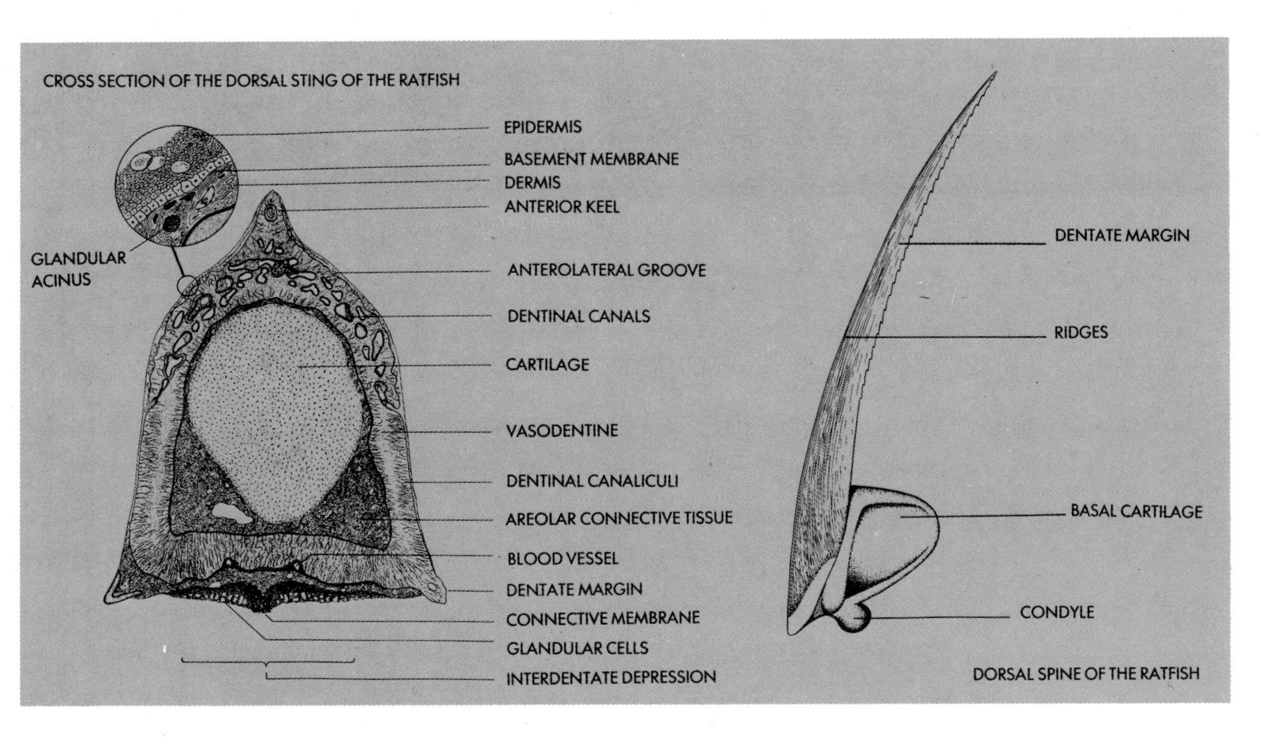

CATFISHES

Family ARIIDAE

There are estimated to be about 1000 species of catfishes, of which most are recorded from fresh water, although there are some from brackish and sea water. A few representative species will be described.

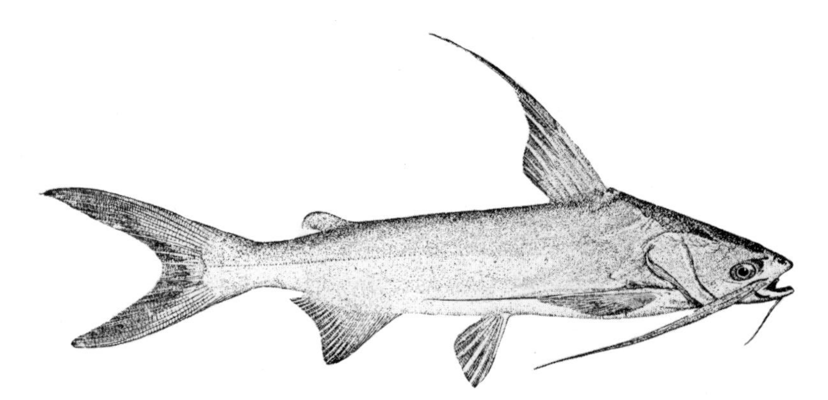

Bagre marinus **(Mitchill)**
Sea catfish or gafftopsail sea catfish

DESCRIPTION Length up to 35 cm (14 inches); bluish gray to dark brown on the upper side and lighter below. The elongated body has two pairs of barbels around the mouth. The dorsal and pectoral fins are each equipped with a serrated erectile spine, both of which are venomous.

HABITAT American Atlantic coast from Cape Cod to Brazil. A marine species which enters estuaries and sometimes ranges upstream into fresh water in the tropics; commonly found in estuaries and mangrove-lined lagoons.

OTHER POINTS Sea catfish of the Ariidae family have an interesting habit in which the male catfish keeps up to 50 or more eggs in his mouth for a period of two months. After the eggs are hatched, the young fish remain in the mouth for an additional period of two weeks.

Galeichthys felis **(Linnaeus)**
Sea catfish or hardhead sea catfish

DESCRIPTION Length up to 40 cm (16 inches); brown to dark brown or dark blue above and somewhat whitish below. The head is rounded with three pairs of barbels around the mouth.

HABITAT Generally found in large shoals on muddy bottoms in turbid waters, usually on the coastline and in river estuaries along the American Atlantic coast from Cape Cod to Panama.

OTHER POINTS The pectoral and dorsal spines are very sharp and make the venom apparatus a formidable and efficient defensive weapon which can inflict an extremely painful wound.

Family CLARIIDAE

These are the walking catfishes, which predominantly inhabit fresh water, with a few species entering brackish estuary areas. This family includes 13 genera with about 100 species.

Clarias batrachus **(Linnaeus)**
Labyrinthic catfish

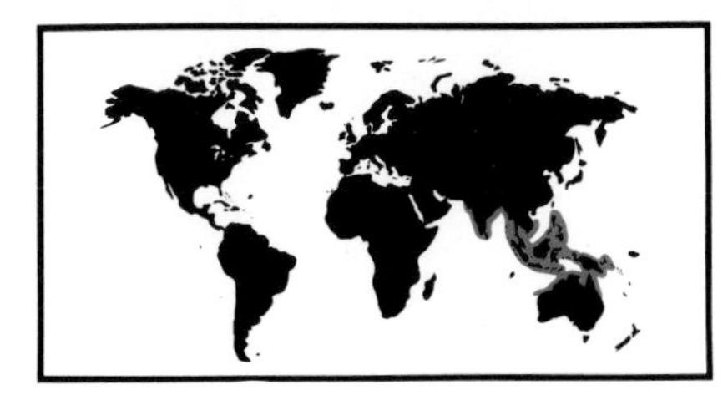

DESCRIPTION Length up to 30 cm (12 inches); dingy green or brownish on the upper side, lighter underneath. The body is long and slightly resembles the shape of an eel. The dorsal fin base is very long, with usually more than 30 rays. These "air-breathing catfishes" commonly have four pairs of barbels dangling from the mouth and head area.

HABITAT Found in fresh and brackish water throughout Africa and most of southern Asia.

OTHER POINTS Differs from other catfishes by having an accessory breathing apparatus situated in a pocket that extends back and upward from the gill cavity. The dendritic breathing organs are attached to the second and fourth gill arches and permit them to live out of water much longer than can most other catfishes. Can move short distances over land. One species of *Clarias* was introduced into southern Florida, where it has become established.

Family PLOTOSIDAE

This family, known as the catfish eels, consists of seven genera and 30 species. The plotosids are largely marine fishes, although a few species inhabit fresh water.

Plotosus lineatus **(Thunberg)**
Oriental or striped catfish

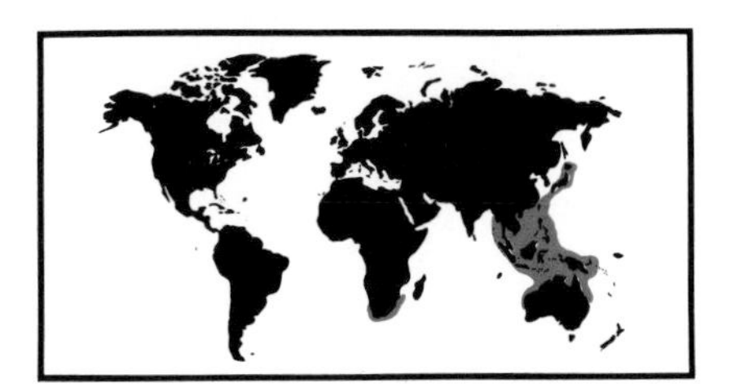

DESCRIPTION Length 30 cm (12 inches); brown or black on the upper side and whitish below with two or three horizontal white or yellow stripes. An eel-shaped elongated body with a dorsal fin extending from the front along the entire body.

HABITAT Very common marine species, also frequenting estuaries and rivers of the Indo-Pacific region. As these fish move about in large schools, they sometimes appear as a dark object several feet in diameter slowly revolving in the water.

OTHER POINTS One of the more dangerous venomous fishes. The dorsal and pectoral spines are very sharp and can inflict an extremely painful wound. Envenomation is rarely fatal.

VENOM APPARATUS OF CATFISHES

Venomous catfishes have a single, sharp and stout spine immediately in front of the soft-rayed portion of the dorsal and pectoral fins. This spine is covered by a thin integumentary sheath. There is no external sign of the venom glands, which are located in a series of sharp, recurving teeth capable of cutting into a victim's flesh, helping the venom to be absorbed and often seeding serious infections. The spines of the catfish are very dangerous once they have been locked into the extended position. For medical advice see Treatment of venomous fish stings, page 118.

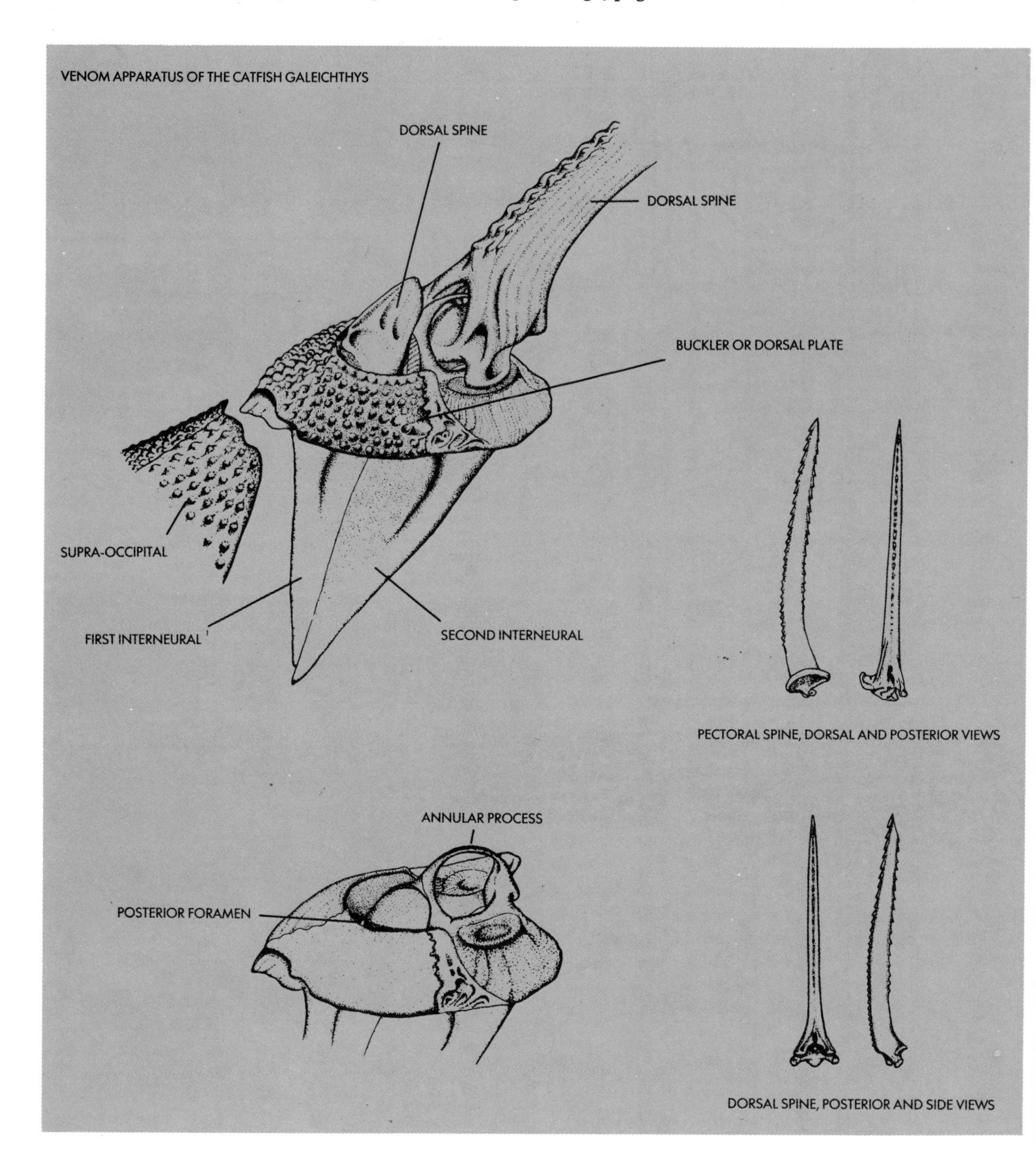

MORAY EELS

Family MURAENIDAE

This family consists of 12 genera with about 100 species. They are typically eel-shaped fishes with long scaleless bodies. Their common habit is to protrude from crevices with a characteristic "gape" of sharp teeth. While they are recorded by some writers as being venomous, no true venom apparatus has been identified; it may be the high frequency of secondary infection that has resulted in the belief that moray bites are venomous. Certainly, the sharp teeth can inflict a painful bite.

Muraena helena Linnaeus
Mediterranean moray eel

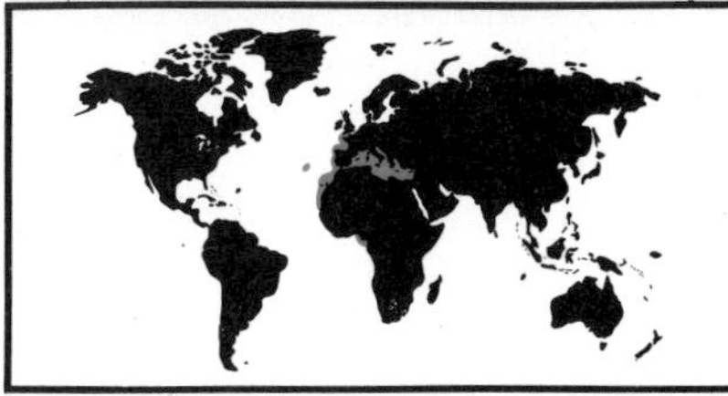

DESCRIPTION Length up to 1 m (39 inches); dark brown or black with yellow spots in an irregular dotted pattern. The body is scaleless, elongated, rounded, and laterally compressed. No pectoral fins. The gill openings are present as a black spot on each side at the jaw level.

HABITAT Usually live relatively close to shore and are found hidden among rocks, crevices and caves in the eastern Atlantic Ocean and Mediterranean Sea.

OTHER POINTS Since ancient times it has been described as possessing hollow fangs and fully developed venom glands, but modern anatomical research does not corroborate this view. The canine teeth are solid, and there is no evidence of venom glands of the types that are found in venomous snakes and other fishes. According to some investigations, the palatine mucosa secretes a toxic substance.

WEEVERFISHES

Family TRACHINIDAE

The weeverfishes (or weevers) consist of one genus, *Trachinus*, with four species, confined to the north-eastern Atlantic and Mediterranean coasts. Despite their relatively small size, generally less than 45 cm (18 inches), all contain venomous dorsal and gill cover spines that are erected when the fish are disturbed. Their habit of lying partly buried in coastal sands makes them a particular danger to persons wading in shallow waters, swimmers and divers. Weevers are caught in commercial trawls and in shrimping nets; great care must be exercised when removing them from the nets.

Trachinus draco **Linnaeus**
Greater weeverfish

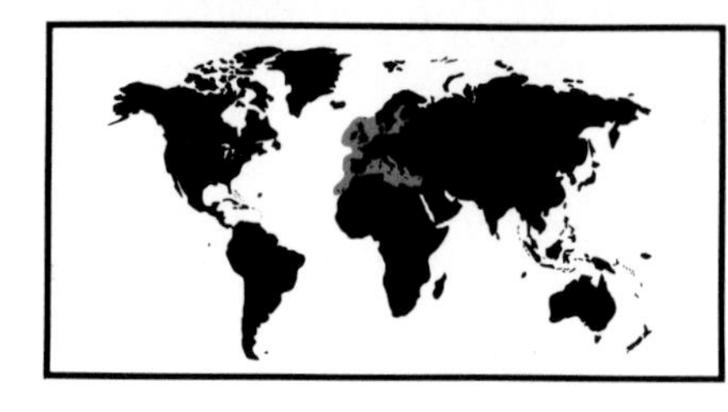

DESCRIPTION Length 45 cm (18 inches); gray-brown on the upper side and yellowish or creamy on the lower sides. There are dark markings along the scales; the anterior dorsal fin is black and contains venomous spines.

HABITAT Common in deeper waters; in the waters of Norway, the British Isles, southward to the Mediterranean Sea, coasts of North Africa and the Black Sea.

Trachinus radiatus **Cuvier**
Weeverfish or starry weever

DESCRIPTION Length 25 to 40 cm (10 to 16 inches); light in color with brown spots and patches. Body is elongated and compressed. There are two dorsal fins; the first short with 6 spines, the second longer with about 25 soft rays. The anal fin is approximately equal in length to the dorsal. The eyes are small and situated on the top of the head.

HABITAT Inhabits sand and mud bottoms of the continental shelf along the coastlines of the Mediterranean Sea, and southward along the west coast of Africa.

***Trachinus vipera* Cuvier**
Lesser weeverfish

DESCRIPTION Length about 14 cm (6 inches); yellow-brown on the back with brown spots, especially on the head; sides and underside are light brown. Thick-set and relatively short-bodied with a compressed head and body. The mouth is large and the eyes are high on the head. There are venom glands on the first dorsal fin, which is totally black, and on the gill cover.

HABITAT Commonly found on clean, sandy bottoms in shallow water areas, where it buries itself in the sand with only the top of its head and back exposed. Found in the North Sea and southward along the coasts of Europe; abundant in the Mediterranean Sea.

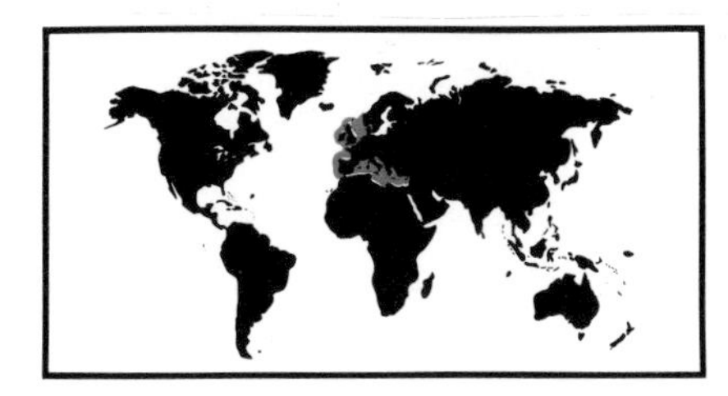

WEEVERFISH STINGS

Venom apparatus of weeverfishes

The venom apparatus of the weeverfish is made up of the dorsal and gill cover spines along with their associated glands. The dorsal spines vary from five to seven in number. Each of the spines is covered by a thin-walled integumentary sheath from which the needle-sharp tip is exposed. Removal of the integumentary sheath reveals a thin, long strip of whitish, spongy venom-producing tissue lying within the grooves, near the tips of each spine. Similar to some snake venoms, weever venom has been found to act both as a neurotoxin and a hemotoxin.

Prevention

Weeverfish stings are most commonly inflicted on persons wading or swimming along sandy coastal areas of the eastern Atlantic Ocean or the Mediterranean Sea. Weevers are usually encountered partially buried in the sand or mud, so persons wading in waters where weevers abound should wear adequate footwear. A living weever should never be handled under any circumstances. Even when dead, weevers can inflict a nasty wound.

Treatment

There is no specific antivenin. The toxin is denatured by heat and a wound should be treated by immersing the affected part in hot water to tolerance (45°C to 113°F). See Treatment of venomous fish stings, page 118.

Above: Buried ***Trachinus vipera***
Left: Drawing showing location of the venom apparatus of ***Trachinus draco***

SCORPIONFISHES

Venomous scorpionfishes have been divided into three main groups on the basis of the structure of their venom organs: zebrafishes – *Pterois* (also called dragonfish or lionfish); scorpionfishes – *Scorpaena*; and stonefishes – *Synanceja*.

Zebrafishes are among the most beautiful and ornate of all coral reef fishes. They are generally found in shallow tropical seas, hovering around crevices or at times swimming gracefully in the open. They are sometimes called turkeyfishes because of their interesting habit of swimming around slowly and spreading fanlike pectorals and lacy dorsal fins like a turkey displaying its plumes. Frequently seen swimming in pairs, they rarely take avoidance action when approached.

Scorpionfishes proper are found from the intertidal zone to depths of 90 m (300 feet), and for the most part live in bays, along sandy beaches, rocky coastlines and coral reefs. Their camouflage coloring and secretive habit of hiding in crevices, among debris, under rocks or in seaweed make them difficult to see. When removed from the water, they erect the spinous dorsal fin and flare the armed gill covers, pectoral, pelvic and anal fins. The pectoral fins are not armed with venom.

Stonefishes are largely shallow water dwellers and are commonly found in tidepools and shallow reef areas. They habitually lie motionless in coral crevices, under rocks, in holes, or buried in the sand or mud. They are very well camouflaged and require extreme agitation to induce movement. They should always be approached with caution.

Family SCORPAENIDAE

This family of scorpionfishes includes 60 genera with over 300 species, the majority of which are found in temperate marine waters. However, most of the venomous members are tropical inhabitants. Members of this family are often extremely difficult to identify, but all have at least one anatomical characteristic in common, which is the presence of a bony plate, or stay, which extends across the cheek from the eye to the gill cover. Consequently, scorpionfishes are sometimes referred to as "mail-cheeked fishes".

Apistus carinatus **(Bloch and Schneider)**
Scorpionfish, bullrout, sulky, waspfish

DESCRIPTION Length commonly 10 cm (4 inches), but up to 18 cm (7 inches). Top of the head has a rough appearance, with three long, slender barbels hanging from the chin. Body shades vary from bluish gray or light yellow above to rosy or white below. The dorsal fin stretches along most of the body, and has a distinguishing black blotch between dorsal spines 8 to 14. The long, winglike black pectoral fin has a free, milky-white ray attached to and hanging from the pectoral base.

HABITAT Soft bottoms, from near shore to a depth of about 60 m (200 feet) in the Indo-Pacific, Australia, Japan, China and India.

OTHER POINTS Has venomous spines.

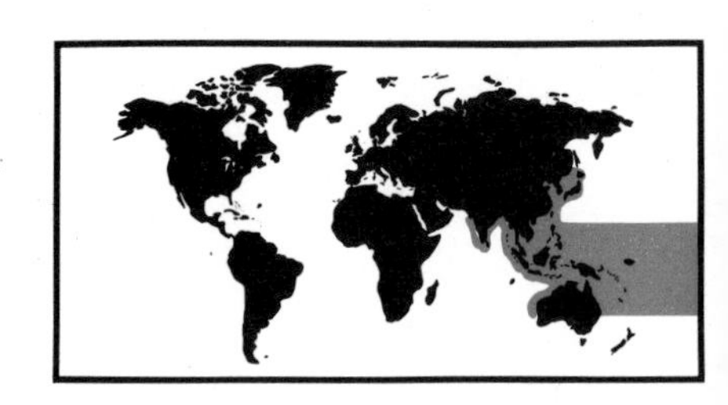

Brachirus (Dendrochirus) brachypterus **(Cuvier)**
Lionfish, turkeyfish, featherfish, short-finned scorpion-fish

DESCRIPTION Length 15 cm (6 inches); short-finned, with alternating dark-brown-to-reddish and light tan vertical bands covering the body. The rows of large spots create conspicuous bands on paired fins.

HABITAT Shallow reef areas of the Indo-Pacific, Australia and the Indian Ocean.

OTHER POINTS The dorsal spines of the members of *Brachirus* are needle-sharp, venomous, and can inflict a very painful wound. The length of the true portions of the pectoral rays varies with the age of the specimen, diminishing in length with age.

Brachirus (Dendrochirus) zebra **(Quoy and Gaimard)**
Lionfish, turkeyfish, featherfish, short-finned scorpion-fish

DESCRIPTION Length 20 cm (8 inches); conspicuous red-orange color with broad white vertical stripes covering the body. The median fins have small dark spots. The cheek has a dark spot and there is usually a banded tentacle above the edge.

HABITAT Shallow water reef areas throughout the Indo-Pacific; Polynesia to east Africa.

Centropogon australis **(White)**
Scorpionfish, waspfish, fortesque, bullrout

DESCRIPTION Length up to 15 cm (6 inches); can vary in color but may have a white body with dark brown head and a brown stripe close to the caudal fin. Has a dorsal fin of proportionately greater height with venomous spines.

HABITAT Shallow coral and rocky areas of New South Wales and Queensland, Australia.

OTHER POINTS Despite its small size, it can inflict a very painful sting.

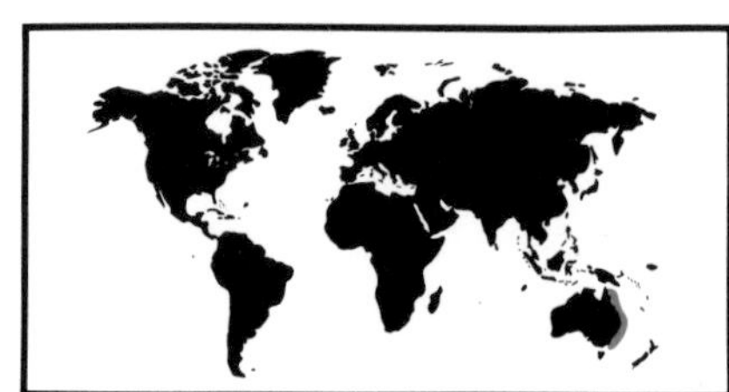

Choridactylus multibarbis **Richardson**
Stonefish

DESCRIPTION Length 10 cm (4 inches); scaleless, with blackish brown fins and a slanted pale band on the dorsal fin. The edges of the pectoral fins are orange, while the inner surfaces are black with several orange bands. The black or dark brown pelvic fins have numerous white spots. The dorsal fin covers the length of the body.

HABITAT Sand or mud bottoms from near shore to a depth of 50 m (165 feet) throughout the Indo-Pacific, India and China.

OTHER POINTS A similar species of scaleless scorpaenid is *C. natalensis*, which has different coloration on the underside of the pectoral fins.

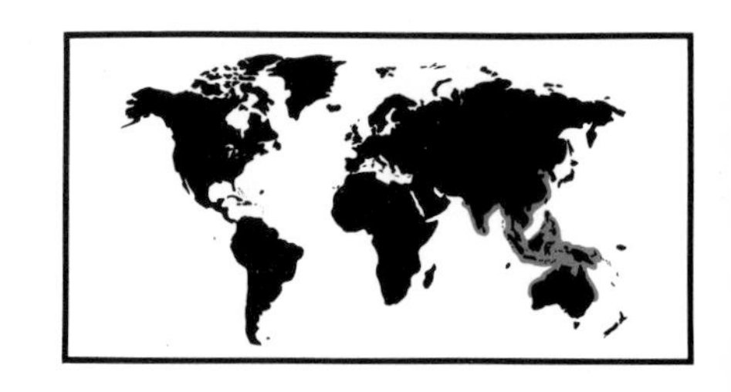

Inimicus didactylus **(Pallas)**
Devil scorpionfish

DESCRIPTION Length up to 13 cm (5 inches); brown body coloration, dotted with white. The dorsal fin has slanted yellow bands, and the pectoral has a wide yellow crossband with yellow spots and band on a black surface on the underside. The caudal fin has two yellow crossbands.

HABITAT A bottom-dwelling fish found in coastal areas throughout the Indo-Pacific.

OTHER POINTS The venom of this fish can be deadly to man. Its ability to camouflage itself by living half-buried presents a real danger.

Inimicus filamentosus **(Cuvier and Valenciennes)**
Devil scorpionfish

DESCRIPTION Length up to 25 cm (10 inches); mostly brown and yellow in body coloration, with brilliant yellow on the inner surface of the pectoral fins. Has extremely elevated and closely set eyes and usually 15 strong, sharp spines which make up the dorsal fin.

HABITAT Sandy bottoms, from near shore to a depth of 55 m (180 feet) throughout the Indo-Pacific.

OTHER POINTS Sometimes confused with *Inimicus sinensis*, which has slightly elevated eye orbits.

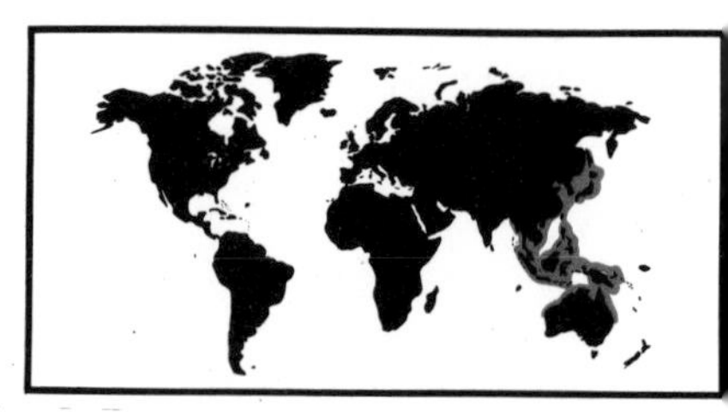

***Minous monodactylus* (Bloch and Schneider)**
Hime-okoze or gray stingfish

DESCRIPTION Length up to 15 cm (6 inches); the upper part of the body has pale bars and stripes, while the underside is usually pale with no markings. The dorsal fin has a large black area or spot, and the edges of the pectoral fins are black. The inner surfaces have no distinctive markings. The head spines are well developed in this scorpaenid and the first dorsal spine is well separated from the remainder of the dorsal fin.

HABITAT Soft bottoms of the continental shelf, from near shore to a depth of 55 m (180 feet) throughout the Indo-Pacific, China and Japan.

OTHER POINTS Members of this genus are subject to considerable variation in color and pattern. They can produce painful stings.

***Notesthes robusta* (Günther)**
Scorpionfish, bullrout

DESCRIPTION Length about 25 cm (10 inches); well-camouflaged, with a dull-yellow background coloration covered by a brown or black mottled design scattered over the entire body, including all fins. The venomous dorsal fin covers about two-thirds of the body length. The pectoral fins are prominent as in most scorpionfishes, but not unusually large.

HABITAT Freshwater streams and brackish shore areas of New South Wales and Queensland, Australia.

OTHER POINTS Because the larger fish of this species are sought by fishermen, it should be remembered that the dorsal fin and the spines at the back of the head can inflict very painful stings.

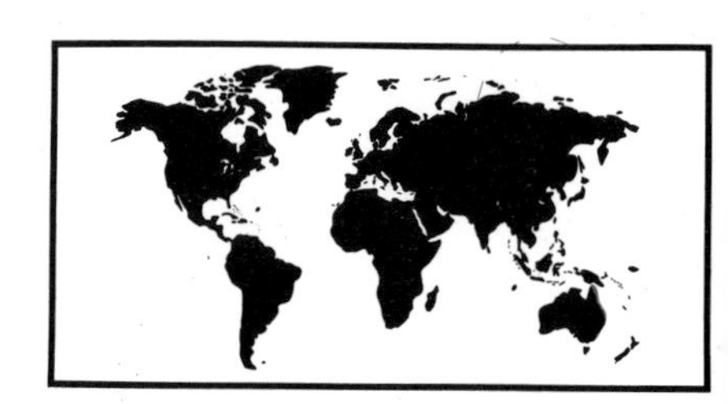

***Pterois antennata* (Bloch)**
Zebrafish, lionfish

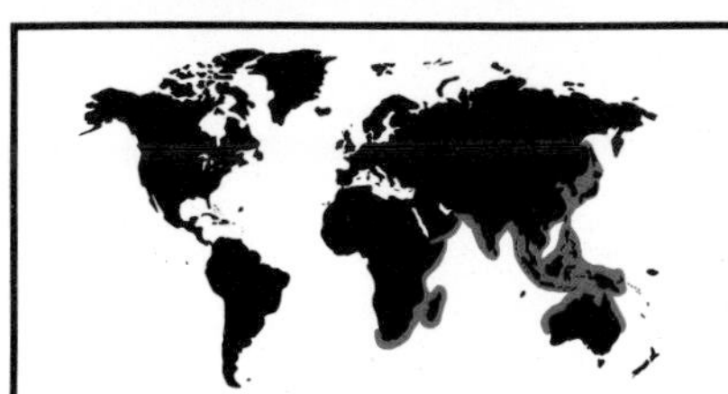

DESCRIPTION Length 20 cm (8 inches); red-orange to tan with white stripes down the body, usually with long dark bands above the eyes. The pectoral fins have scattered dark spots.

HABITAT The crevices and caves of many coral reef areas throughout the Indo-Pacific, Indian Ocean and China.

OTHER POINTS Venomous, and capable of inflicting a painful sting.

***Pterois radiata* Cuvier**
Zebrafish, lionfish, tigerfish, scorpionfish, turkeyfish, firefish

DESCRIPTION Length 20 cm (8 inches). One of the most magnificent species of the group, having long, white and graceful pectoral rays radiating from a reddish to brownish body with white lines circling top to bottom.

HABITAT Usually found on reefs; may be most active in darker hours, staying in caves and crevices during the day. Widely distributed in the Indo-Pacific, Indian Ocean and the Red Sea.

OTHER POINTS Venomous and can inflict painful stings.

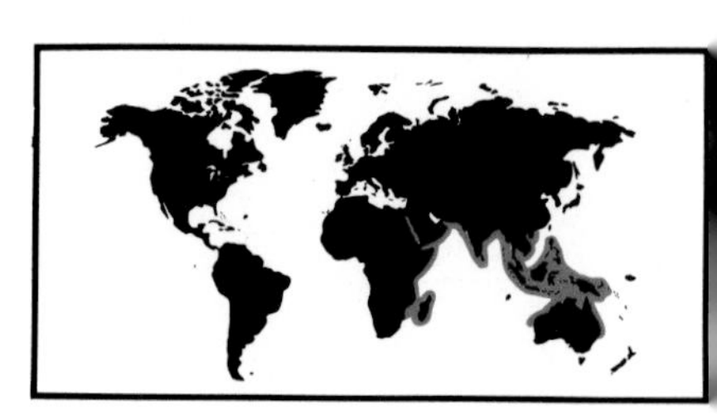

***Pterois russelli* Bennett**
Also known as *Pterois lunulata*
Zebrafish, lionfish

DESCRIPTION Length 30 cm (12 inches); body color varies from red to orange, with many thin dark lines on both the body and head. The large pectoral rays are in a fan-like shape with dark markings along the spines.

HABITAT Throughout the Indo-Pacific, Australia and the Indian Ocean.

OTHER POINTS It is important to remember that hidden within the lacy fins are venomous spines which are dangerous to handle.

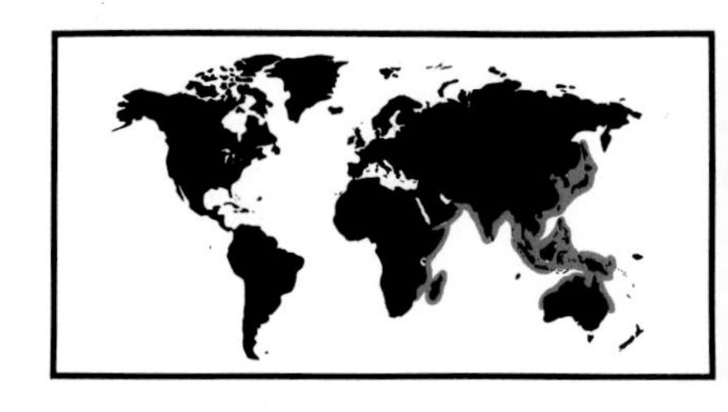

***Pterois volitans* (Linnaeus)**
Zebrafish, lionfish, tigerfish, scorpionfish, turkeyfish

DESCRIPTION Length 38 cm (15 inches); maroon or brownish red, with many dark-red or blackish stripes bordered by white lines. The dorsal, anal, and caudal fins have rows of black spots. Colorful and graceful, with about 12 scales between the median dorsal spines and the lateral line.

HABITAT Generally found in shallow water, hovering in a crevice or swimming leisurely in the open. Distributed in the Indo-Pacific, Australia, Japan, China, Indian Ocean and the Red Sea.

***Scorpaena grandicornis* Cuvier and Valenciennes**
Lionfish, long-horned scorpionfish

DESCRIPTION Length 30 cm (12 inches); brown with white markings. A distinctive-appearing scorpionfish, with long fan-like tendrils standing erect from the head and a taller dorsal fin. The tendril sprouting from the head has given rise to the descriptive name of "long-horned scorpionfish".

HABITAT Seagrass beds, grassy bays and channels throughout the Florida Keys, West Indies, Panama and Brazil.

OTHER POINTS Wounds inflicted are extremely painful, but not fatal.

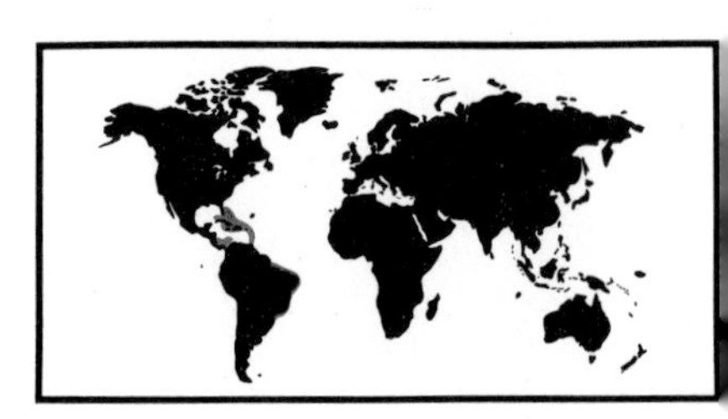

***Scorpaena guttata* Girard**
California scorpionfish, sculpin

DESCRIPTION Length 43 cm (17 inches); body basically red to brown, but specimens from greater depths tend toward red coloration; brown spots and markings cover the body and all fins. A large mouth and large fan-like pectoral fins. The dorsal fin has 12 spines and a row of smaller spines under each eye.

HABITAT Rocky shores and bays in shallow water to a depth of approximately 180 m (600 feet). Often found in caves and crevices along the Central California coast; south into the Gulf of California.

OTHER POINTS Its spines are venomous.

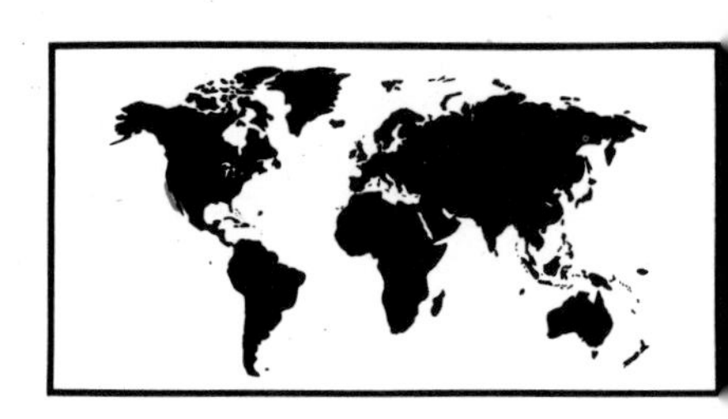

***Scorpaena plumieri* Bloch**
Spotted scorpionfish, sculpin

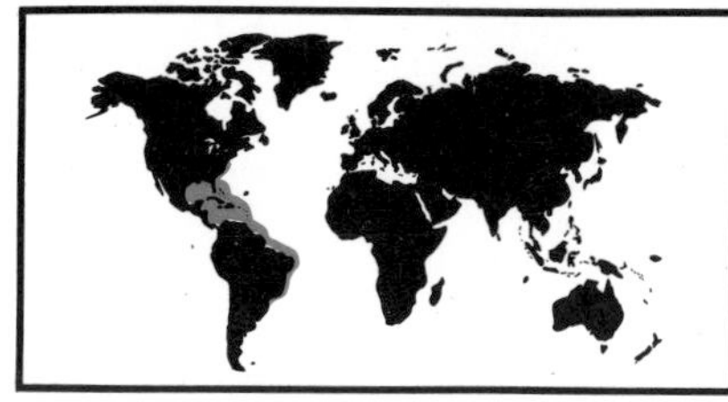

DESCRIPTION Length 30 cm (12 inches); a stout, spine-headed fish that can have brown or black markings on a paler background. The area in front of the caudal fin is decidedly paler, the fins variously blotched, and the head mottled.

HABITAT Warm, shallow coral reef and rocky areas; can range as far as 60 m (200 feet) offshore along the Atlantic coast from Massachusetts to the West Indies and Brazil.

OTHER POINTS Generally quiescent, but when agitated will settle down tightly on the sand and sometimes arch its back. If an intrusion continues, the fish can suddenly change its stance and expose large yellow patches on the pectoral fins. The pectorals can be flipped over, displaying the brightly-colored undersurface. This behavior is somewhat similar to the "eyespot" displays of some moths. Has venomous spines.

Scorpaena porcus **Linnaeus**
Scorpionfish

DESCRIPTION Length about 30 cm (12 inches); well-camouflaged, being brownish-red and dotted with black. Often, half-grown individuals have a black blotch on the posterior half of the spinous dorsal fin, and black dots on the caudal fin arranged in crossbands. Head is scaleless and large.

HABITAT Along the Atlantic coast of Europe from the English Channel to the Canary Islands, French Morocco, the Mediterranean Sea and the Black Sea.

OTHER POINTS The dorsal fin can inflict a painful sting. The pectoral fins, although dangerous in appearance, are not venomous.

Scorpaena scrofa **Linnaeus**
Scorpionfish

DESCRIPTION Length 50 cm (20 inches); the body has dark blotches on a light background, with a dark spot present on the dorsal spine. The eyes face upwards and there are small branch-like growths protruding from the head.

HABITAT Bays, sandy beaches, rocky coastlines or coral reefs along the west coast of France south to Cabo Blanco (Argentina), northwest Africa and the Mediterranean Sea.

OTHER POINTS One of the species of venomous fishes best known to European writers of antiquity.

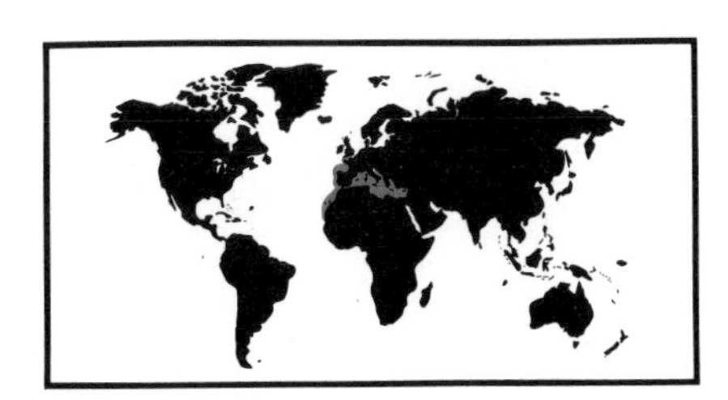

Scorpaenopsis diabolus **(Cuvier)**
Scorpionfish

DESCRIPTION Length 25 cm (10 inches); a solid-appearing scorpionfish of darker coloration; it can be red, marbled with brown and bluish-white, with the pectorals spotted or banded with black on the undersides. The caudal fin has brown crossbands.

HABITAT Shallow coral and rocky areas of the Indo-Pacific and Australia.

OTHER POINTS Can inflict a painful injury with its venomous dorsal sting.

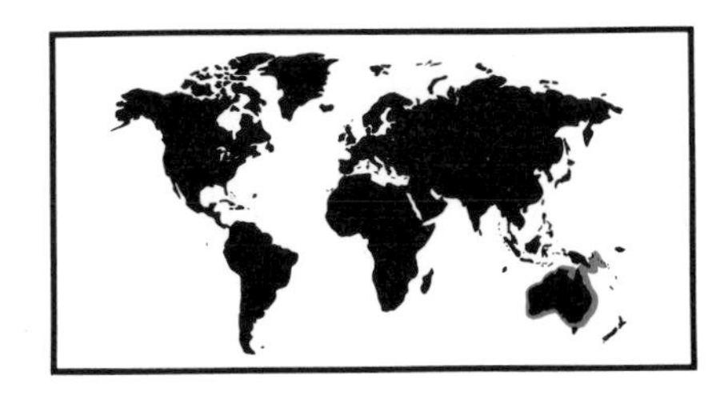

***Scorpaenopsis gibbosa* (Bloch and Schneider)**
Scorpionfish

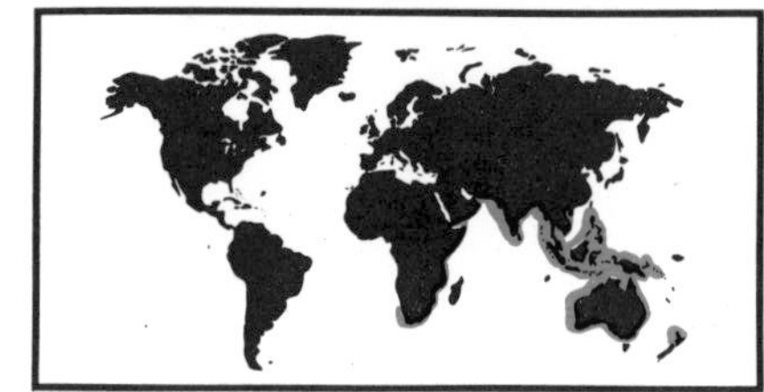

DESCRIPTION Length 20 cm (8 inches); the head, body and fins are camouflaged with green fleshy appendages. The overall body coloring is often mottled with greenish-gray, while some specimens may be dull red with orange-tipped fins. The caudal fin has two broad black stripes. Distinguishable from the stonefish in that it has brightly colored golden-yellow and black roundish blotches on the underside of its pectoral fins. The dorsal fin is of medium height, the pectoral fins are large.

HABITAT Bottom-dweller living next to the rims of coral reefs throughout the Indo-Pacific and Indian Ocean.

OTHER POINTS Can inflict painful stings.

***Sebastes caurinus* Richardson**
Copper rockfish

DESCRIPTION Length 60 cm (24 inches); a thick-bodied fish which can be of varied color. It is commonly dark brown or olive to pink or orange-red on the topside, with copper-pink or yellow markings, while the underside is whitish. Often there is a whitish stripe along the lateral line.

HABITAT Rocky areas or rocky and sandy bottom areas in shallow water from the Gulf of Alaska to central Baja California, and the Sea of Japan.

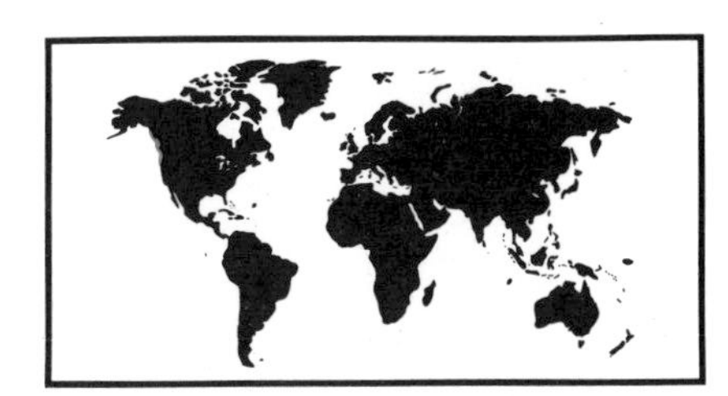

***Sebastes dallii* (Eigenmann and Beeson)**
Calico rockfish

DESCRIPTION Length 25 cm (10 inches), but usually smaller; slanting red-brown bars, along with scattered brownish blotches covering a yellow-green body color. The caudal fin has brown or red-brown streaks and spots.

HABITAT Found near soft bottom areas between central California and Baja California.

OTHER POINTS The color pattern is unique among rockfish. Its spines are venomous.

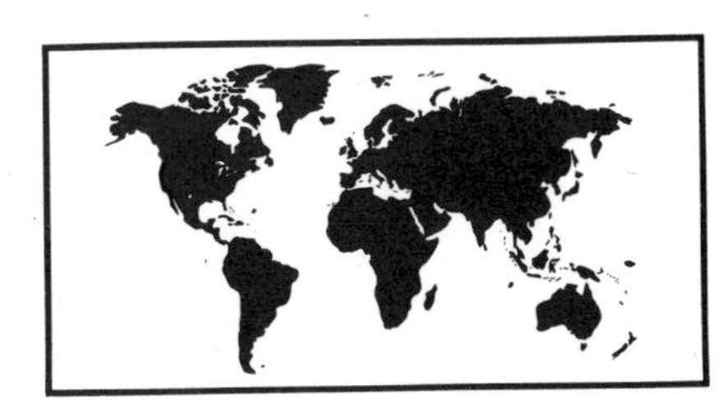

***Sebastes nebulosus* Ayers**
China rockfish

DESCRIPTION Length about 43 cm (17 inches); body coloration can be black or blue-black with mottled yellow markings. There is a distinctive yellow line running along the side of the body from the 3rd or 4th dorsal fin to the caudal fin.

HABITAT An inshore species, which lives along the open coast among rocks and reefs from Alaska to southern California.

OTHER POINTS A prize catch of many fishermen, due to its reported tastiness. Has venomous spines.

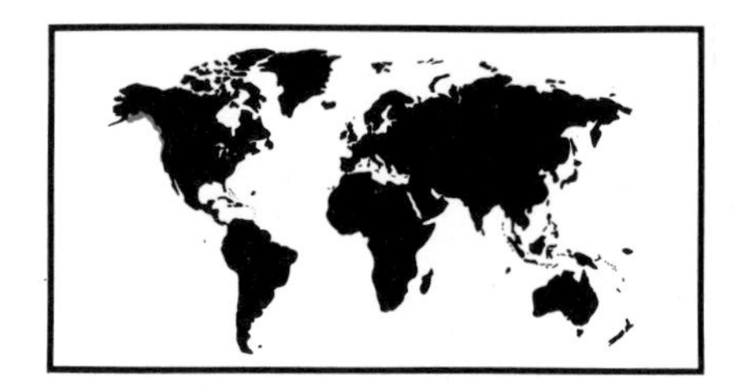

***Sebastes paucispinis* Ayers**
Bocaccio

DESCRIPTION Length 90 cm (36 inches); brownish in color, but may be pink on the underside. All fins are brownish. Has a large mouth and prominent lower jaw.

HABITAT Rocky reefs and open bottom from the Gulf of Alaska to central Baja California.

OTHER POINTS An important commercial fish along the California coast. Its spines are venomous.

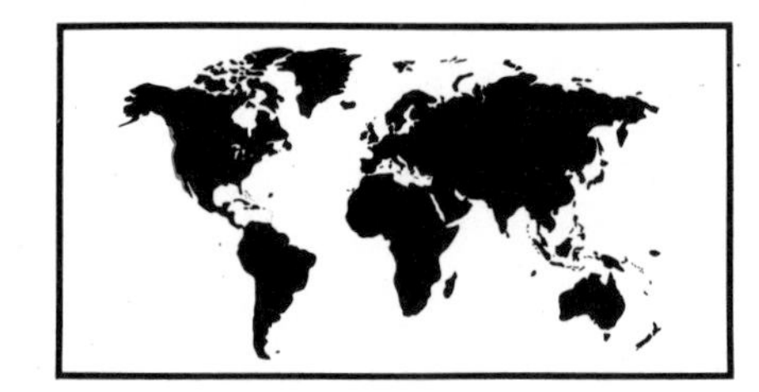

***Sebastes pinniger* (Gill)**
Canary rockfish

DESCRIPTION Length up to 76 cm (30 inches); gray-orange in color, with bright orange fins. The lateral line has a gray area, except in some larger adults.

HABITAT Rocky bottom areas from Alaska to northern Baja California.

OTHER POINTS The rear of the dorsal fin in smaller specimens may be a dusky color. The spines are venomous.

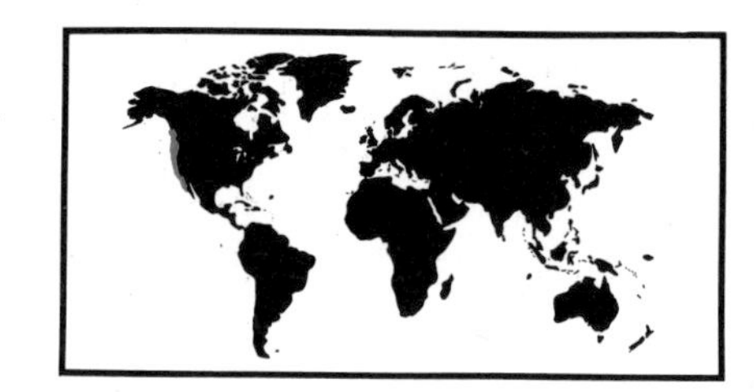

***Sebastes rosaceus* Girard**
Rosy rockfish

DESCRIPTION Length about 36 cm (14 inches); the white blotches bordered by purple on the back stand out from the red, with some yellow coloration of the overall body. The fins are orange-red and there are often purple stripes on the head.

HABITAT A bottom-dweller at around 30 to 46 m (100 to 150 feet), commonly found from northern California to central Baja California.

OTHER POINTS Often confused with the Rosethorn rockfish, *Sebastes helvomaculatus*, which has white blotches bordered by red on the back. Has venomous spines.

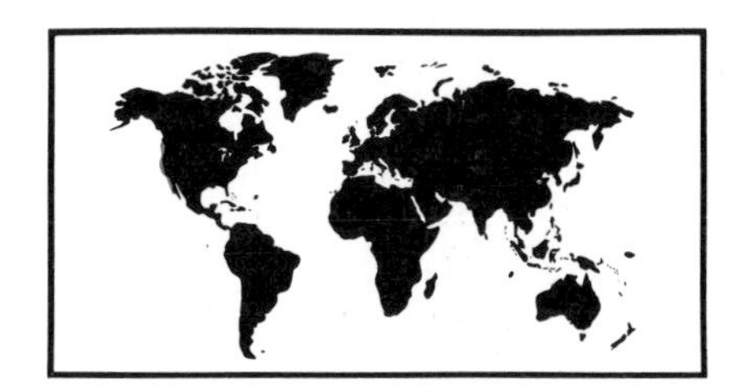

***Sebastes serriceps* (Jordan and Gilbert)**
Treefish

DESCRIPTION Length 41 cm (16 inches); yellow to olive, with thick black vertical stripes and pink lips. All fins have the same coloration as the body.

HABITAT From northern California to central Baja California, in rocky areas, usually in crevices.

OTHER POINTS A territorial species with venomous spines.

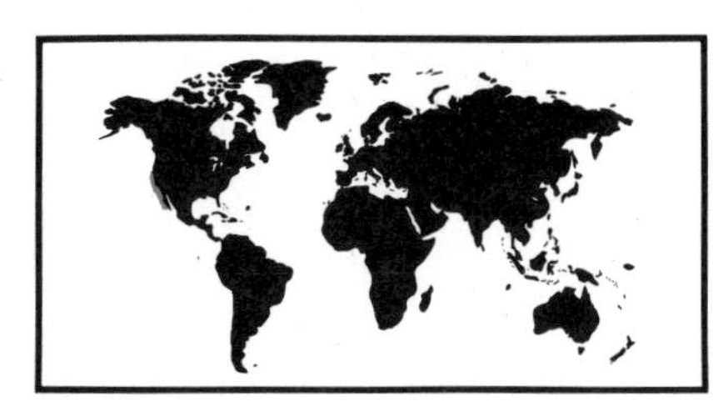

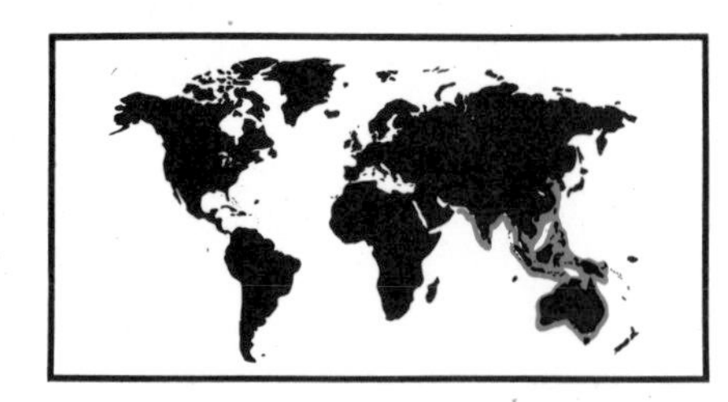

Synanceja horrida **(Linnaeus)**
Stonefish

DESCRIPTION Length 60 cm (24 inches). Because of a thick coating of slime, irregular wartlike texture of skin, and habit of burying itself in the sand, it becomes coated with bits of coral debris, mud and algae. Specimens have been observed with algae growing on their skin. An unattractive fish, a sluggish swimmer with small eyes and 13 strong dorsal spines. The pectoral fins have 16 rays. The mouth cavity is yellowish-green.

HABITAT Makes a shallow depression by scooping up sand or mud with its pectoral fins until it is piled up around the sides of its body. Lies motionless on sandy bottom areas throughout the Indo-Pacific, Australia, China and India.

OTHER POINTS Undoubtedly one of the most dangerous venomous species. The venom glands are the largest found in fishes, causing extremely painful stings and human fatalities.

Synanceja verrucosa **Bloch and Schneider**
Stonefish

DESCRIPTION Length 30 cm (12 inches); the brownish body is without scales and the head is strongly depressed with the eyes only slightly elevated and far apart. The pectoral pelvic and caudal fins are all tipped with white.

HABITAT Shallow waters among coral reefs and coral rubble and in pools at low tide throughout the Indo-Pacific, Australia, Indian Ocean and the Red Sea.

OTHER POINTS The stonefishes *Synanceja* are extremely dangerous. Because of their camouflaged appearance, they frequently resemble a large clump of mud or debris. Their stings are excruciatingly painful and occasionally cause human deaths.

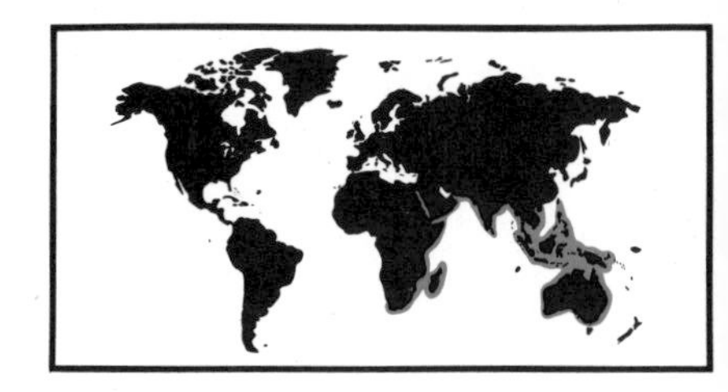

VENOM APPARATUS OF SCORPIONFISHES

The zebrafish has a venom apparatus consisting of 13 dorsal spines, three anal spines, two pelvic spines, and the associated venom glands. The spines are for the most part long, straight, slender, and camouflaged in delicate, lacy-appearing fins. Located on the front side of each spine are the glandular grooves, open on either side, which appear as deep channels running along the entire length of the shaft. Situated in these grooves are the venom glands. Each gland is totally covered by the integumentary sheath.

The venom apparatus of the scorpionfish proper includes 12 dorsal spines, three anal spines, two pelvic spines, and their venom glands. The spines are shorter and heavier than those found in the zebrafish type. The glandular grooves are only located in the upper two-thirds of the spine. The venom glands lie along these grooves and are limited to the upper half of the spine. The integumentary sheath is moderately thick.

The stonefish usually has 13 dorsal spines, three anal spines, two pelvic spines, and their associated venom glands. The venom organs of this fish are different from the other types because of the short, heavy spines and greatly enlarged venom glands, covered by a very thick layer of warty skin. The pain resulting from a sting from this dangerous fish is immediate, intense, can last for a number of days and may be fatal. For medical advice see Treatment of venomous fish stings, page 118.

Spines of the stonefish ***Synanceja horrida***

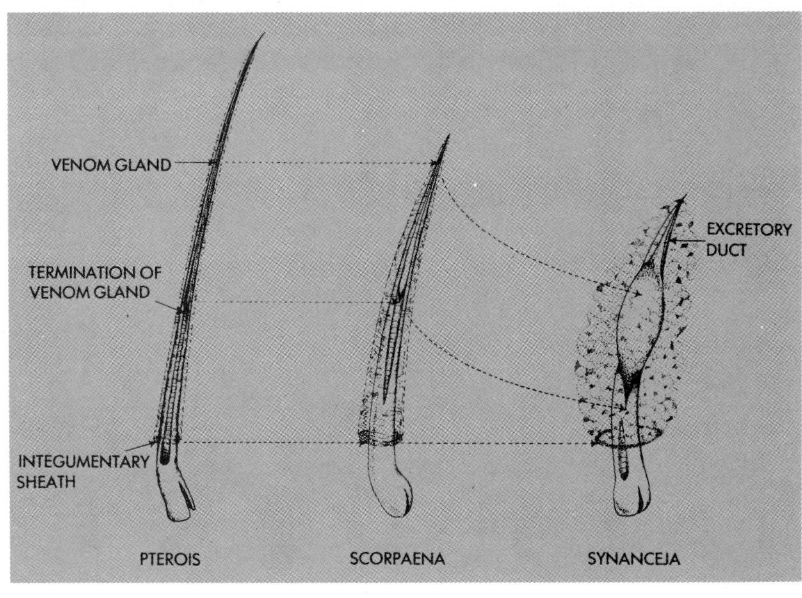

Comparison of the three types of scorpionfish dorsal sting

TOADFISHES

Toadfishes are small bottom-dwelling fishes found in warmer waters along the coasts of America, Europe, Africa and India. Most of them live in the oceans and seas, but some live around river mouths or entirely in fresh water, being found at great distances up-river. They hide in crevices, burrow under rocks, debris, or among seaweed, or lie almost completely buried under a few centimeters of sand or mud. Toadfishes tend to migrate to deeper water during the winter months, when they remain in a motionless state of hibernation. They are experts at camouflage and have the ability to change to lighter or darker shades of color. Combined with patterns of markings and blotches of different colors, this often renders these particular fishes difficult to identify in their normal habitat.

Family BATRACHOIDIDAE

This is a family of small to medium sized bottom-dwelling and primarily marine fishes consisting of approximately 22 genera and about 55 species.

Batrachoides didactylus **(Bloch)**
Toadfish

DESCRIPTION Length 20 cm (8 inches); the body is spotted all over with brown and the spotting is separated by whitish lines. The dorsal fin carries larger spotting. The eyes are very small, and the snout is very broad, flat and depressed, while surrounded by tentacles of various length.

HABITAT Bottom areas in the Mediterranean Sea and nearby Atlantic coasts.

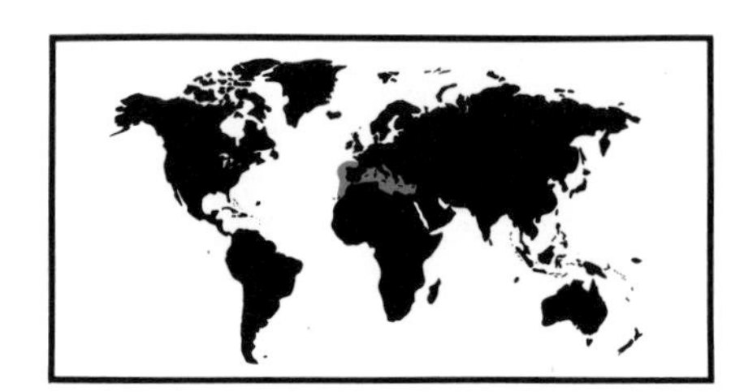

Batrachoides grunniens **(Linnaeus)**
Grunting toadfish or munda

DESCRIPTION Length 20 cm (8 inches); the head, body and dorsal fins are brown with marbled light and dark areas. The snout is broad, flat, depressed and surrounded by a circle of short tentacles.

HABITAT Bottom areas mainly in the coastal waters of Sri Lanka, India, Burma and Malaysia.

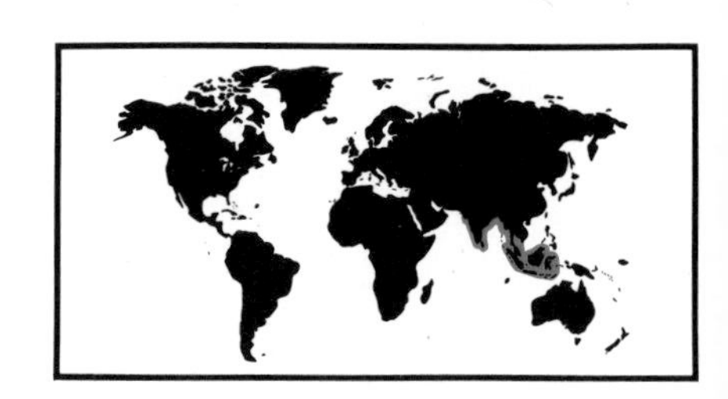

***Opsanus tau* (Linnaeus)**
Oyster toadfish

DESCRIPTION Length more than 30 cm (12 inches); well camouflaged dark-green to brownish mottled coloration on body. Alternating slanted lines on dorsal fin and vertical dark stripes on caudal fin. Pectoral fins are a mottled coloration similar to the body.

HABITAT Shallow water areas among weeds and debris along the North Atlantic coast from Maine to the Caribbean.

OTHER POINTS Slow-moving in its habits.

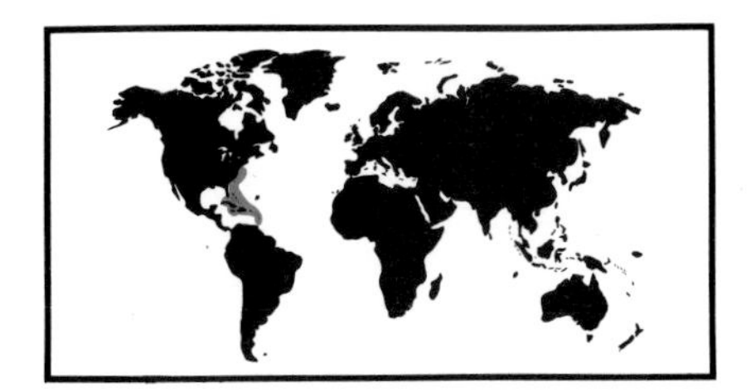

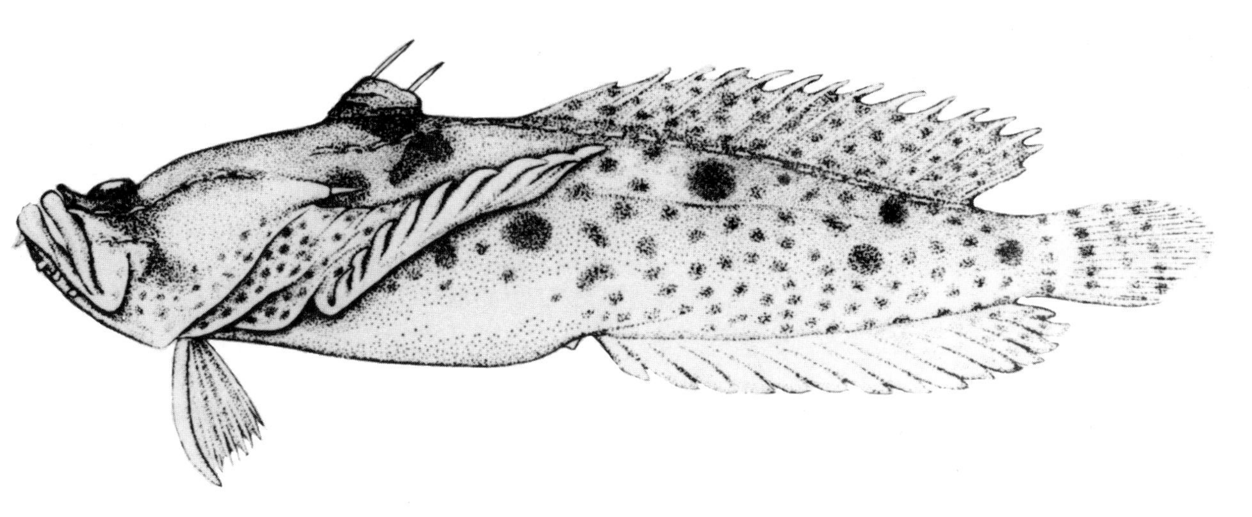

***Thalassophryne maculosa* Günther**
Toadfish

DESCRIPTION Length up to 14 cm (5.5 inches); the scaleless body tends to be light, with small to medium dark spots and blotches in smaller specimens, while the larger examples have a darker background. The first two dorsal fins are set distinctly apart and connected to venom glands. The dorsal, pectoral, and caudal fins are pigmented to the distal margins. There are relatively few flat barbels on the chin.

HABITAT Partially buried in mud or sand throughout the Caribbean Sea.

OTHER POINTS Can inflict painful wounds with its venomous dorsal and opercular spines.

VENOM APPARATUS OF TOADFISHES

The venom apparatus of toadfishes includes two dorsal fin spines and two gill cover spines, along with their venom glands. The dorsal spines are slender and hollow, slightly curved, and end in sharp, needle-like points. At the base and tip of each spine is an opening through which the venom passes. The base of each dorsal spine is surrounded by a gland-like mass in which the venom is produced. Each gland empties into the base of one of the spines. The bony covering protecting the gills of the fish, the operculum, is also used as a defensive venom organ. For medical advice see Treatment of venomous fish stings, page 118.

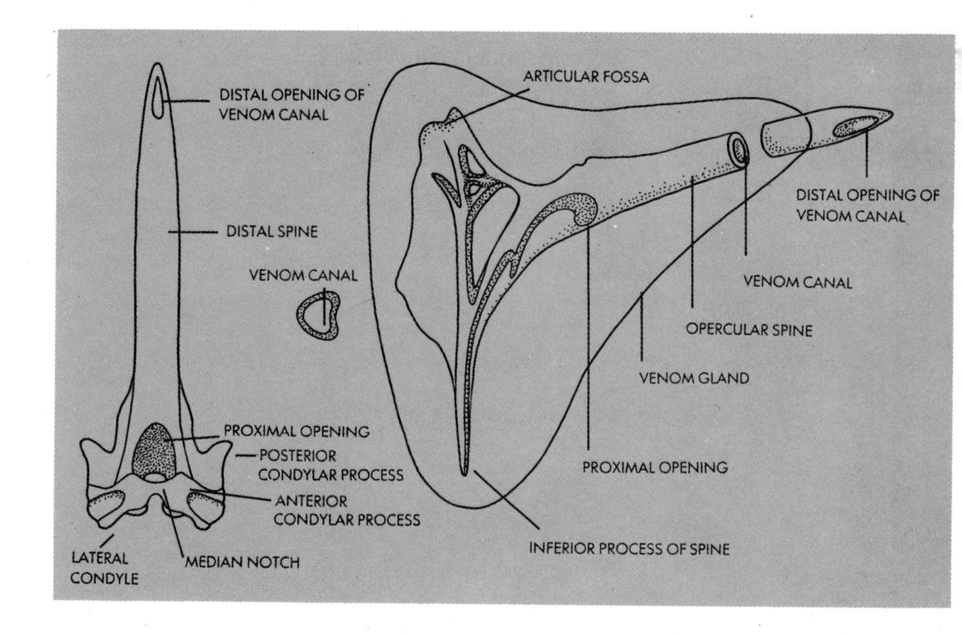

Anatomy of the venom apparatus of the toadfish ***Thalassophryne dowi*** *(right) with the dorsal sting shown far right*

Dorsal sting of the toadfish

SURGEONFISHES

Surgeonfishes are members of the family Acanthuridae and live in reefs and warm seas. They are herbivorous fishes, the family consisting of approximately 77 species in 10 genera. The largest genus is characterized by a sharp, lancelike, and moveable spine on the side and base of the tail fin. Normally the spines lie flat against the body within deep recesses, partially covered by a layer of skin. When the fish becomes excited, the spine can be pointed forward, making a right angle with the body. With a quick, lashing movement of the tail and spine, large surgeonfishes are capable of making deep and painful wounds. Some species appear to have venomous spines, while others do not. All, however, should be handled with care.

Family ACANTHURIDAE

Acanthurus achilles Shaw
Achilles surgeonfish

DESCRIPTION Length about 20 cm (8 inches); background body coloration is dark brown to black. Teardrop-shaped body with a large orange-red spot on the side where the spine is situated. There is also a red area on the tail and a long, narrow, red line at the base of both the dorsal and anal fins. There is a light blue line under the chin and another on the edge of the operculum under the eye.

HABITAT Inshore areas near coral reefs throughout surge channels, from Hawaii to central Polynesia, and Micronesia to Melanesia.

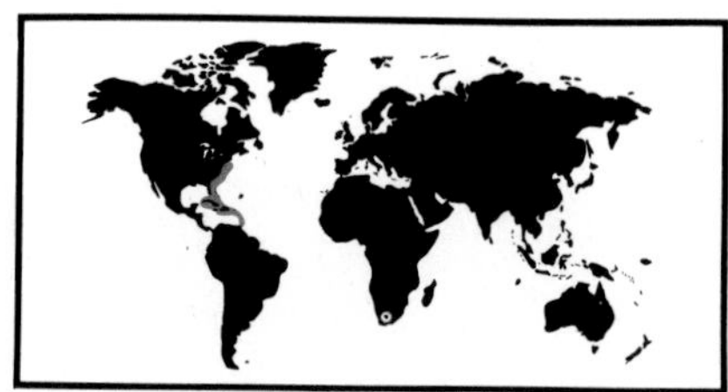

Acanthurus coeruleus **Bloch and Schneider**
Blue tang

DESCRIPTION Length 35 cm (14 inches); a high-bodied, compressed fish which is purplish gray to blue with narrow gray bands running longitudinally along the body. The juvenile coloration may be bright yellow. The continuous unnotched dorsal fin has 9 spines and 26 to 28 soft rays.

HABITAT Warm seas; particularly common in surge channels and shoal areas near coral patch reefs throughout the western Atlantic from New York to the West Indies.

VENOM APPARATUS OF SURGEONFISHES

Some species have retractable spines on the side at the base of the tail, which can both envenom and lacerate. Care should be taken when removing the fish from a hook or net. For medical advice see the section on Treatment of venomous fish stings, page 118.

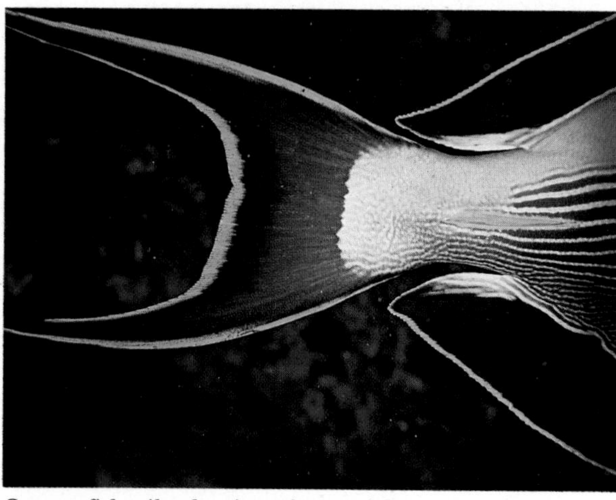

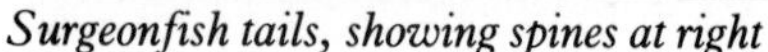

Surgeonfish tails, showing spines at right

DRAGONETS

Dragonets are small, scaleless fish with flat heads and a spiny opercular covering over the gills. The spines are extremely sharp and can easily penetrate the skin. The fish show striking sexual dimorphism, especially in the breeding season, when males show bright iridescent colors and long dorsal fin filaments. Some species are found in deep water, others in shallow inshore bays and reefs.

Family CALLIONYMIDAE

Callionymus lyra Linnaeus
Dragonet

DESCRIPTION Length 10 cm (4 inches). The adult male is yellowish and spotted with lilac. It is brightly variegated in color and has a high and often filamentous dorsal fin. A small, scaleless fish with a flat head. The female is less colorful, with a scarcely prolonged dorsal spine.

HABITAT Shores along the Atlantic coast of Europe and the Mediterranean Sea.

OTHER POINTS Said to be venomous, but this is not documented.

RABBITFISHES

Rabbitfishes are a group of herbivorous, spiny-rayed fishes with an appearance similar to surgeonfishes. They differ from other fishes because the first and last rays of the pelvic fins are modified into sharp spines. Rabbitfishes are of moderate size, and are plentiful around rocks and reefs from the Red Sea to Polynesia. This family is made up of two genera with about 10 species.

Family SIGANIDAE

Siganus puellus **(Schlegel)**
Rabbitfish

DESCRIPTION Length 27 cm (10.5 inches); oval-shaped body covered by vivid shades of yellow and orange. All fins are yellowish in coloration. Has an interesting mask-like stripe from the chin to the front of the first dorsal spine.

HABITAT Among rocks and reef areas throughout the Indo-Pacific.

OTHER POINTS Can inflict painful stings.

Siganus spinus **Linnaeus**
Rabbitfish

DESCRIPTION Length up to 25 cm (10 inches); a fairly slender, compressed body, the head and trunk covered with pearly blue to cream patterning. There are also various shades of brown or gray. The pattern of coloration of the body extends onto the dorsal and anal fins.

HABITAT Shallow coral reef flats in small schools throughout the Indo-Pacific and the Red Sea.

OTHER POINTS Can inflict painful stings.

Siganus vulpinus **(Schlegel and Müller)**
Rabbitfish

DESCRIPTION Length up to 20 cm (8 inches); oval-shaped body outline with orange coloration and a dark brown band on the face from the snout to the first dorsal spine. There are a great number of brown dots on the upper portion of the body. The dorsal fin covers most of the length.

HABITAT The Indo-Pacific area.

OTHER POINTS Capable of inflicting painful stings.

VENOM APPARATUS OF RABBITFISHES

The venom apparatus of the rabbitfish is made up of 13 dorsal, four pelvic and seven anal spines, along with their venom glands. There is a groove along both sides of the mid-line of the spine for almost the entire length. These grooves are generally deep and contain the venom glands, which are located in the outer one-third of the spines near the tips. For medical advice see Treatment of venomous fish stings, page 118.

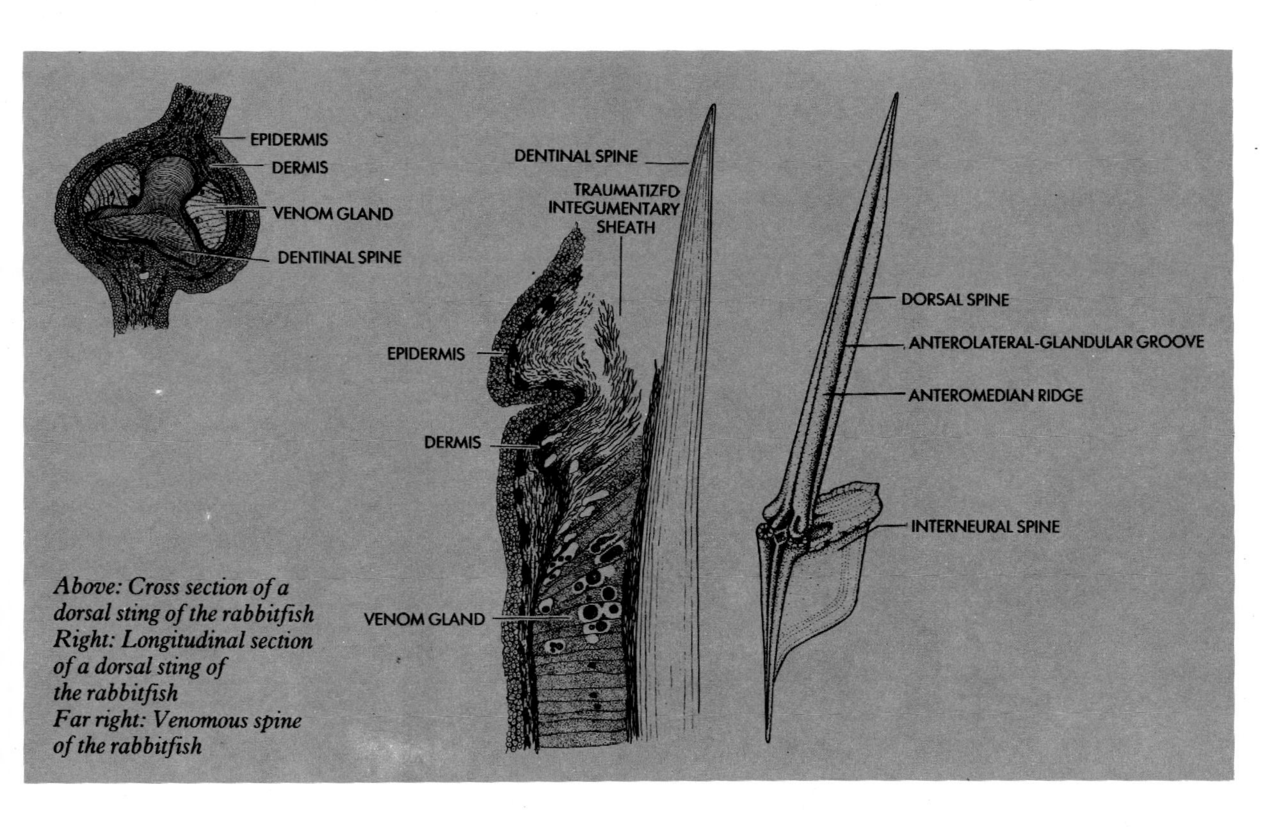

Above: Cross section of a dorsal sting of the rabbitfish
Right: Longitudinal section of a dorsal sting of the rabbitfish
Far right: Venomous spine of the rabbitfish

STARGAZERS

Family URANOSCOPIDAE

Stargazers are bottom-dwelling marine fishes with cube-shaped heads, nearly vertical mouths with fringed lips, and eyes on the flat upper surface of the heads. Uranoscopids spend a large part of their time buried in the mud or sand, with only the eyes and a portion of the mouth visible. There are eight genera, which include about 25 species.

Uranoscopus scaber **Linnaeus**
Stargazer

DESCRIPTION Length 15 cm (6 inches); the back and sides are grayish brown, speckled with white and yellowish white on the underside. The first dorsal fin is black. The front of the body is massive, with small eyes on top of the head and an almost vertical mouth.

HABITAT Usually found buried in the sand or mud throughout the eastern Atlantic and the Mediterranean.

OTHER POINTS Its shoulder spines can inflict painful stings.

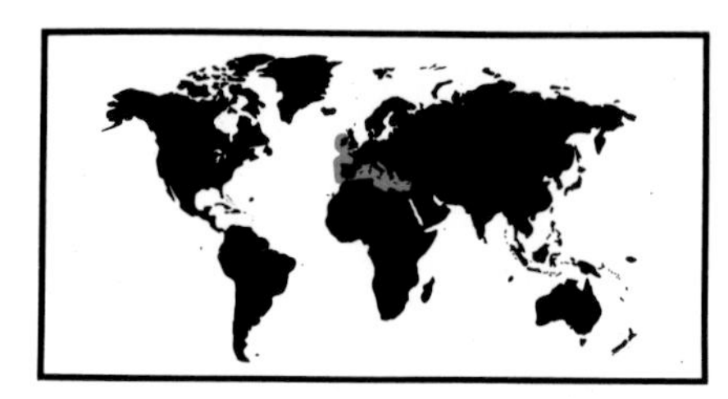

VENOM APPARATUS OF STARGAZERS

The venom apparatus consists of two shoulder spines, one on either side, which stick out through a layer of skin. Venom glands are attached to these spines. The spine is said to have a double groove, through which the venom flows. For medical advice see Treatment of venomous fish stings, page 118.

HEAD OF URANOSCOPUS SCABER SHOWING SHOULDER SPINES

LEATHERBACKS

Family CARANGIDAE

Leatherbacks are members of the family Carangidae, which also includes jacks, scads and pompanos. The family is made up of approximately 24 genera, with about 200 species. A particular characteristic of this family is the presence of two separate spines in front of the anal fins.

Several carangids are believed to have venomous spines, but the venom apparatus has been described in only *Scomberoides sanctipetri* (Cuvier), which inhabits the tropical Indo-Pacific region. The carangids are a group of fast-swimming oceanic fishes which are usually found around coral reefs and islands and are of considerable commercial importance.

Scomberoides sanctipetri **(Cuvier)**
Leatherback

DESCRIPTION Length 65 cm (25 inches); body coloration is gray-green on the dorsal portion, with silver-gray to the midline and silvery-white on the underside. Oblong body shape, with the dorsal and ventral fins almost equal in length. There is a forked caudal fin, and the dorsal fins are dark at the tips.

HABITAT Lives within small schools and inhabits waters from shallow lagoons to offshore areas throughout the Indo-Pacific region.

OTHER POINTS Can inflict painful stings.

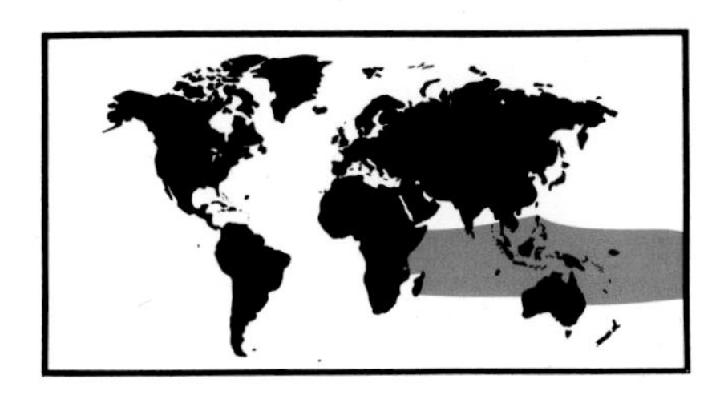

VENOM APPARATUS OF LEATHERBACKS

The venom apparatus consists of seven dorsal spines and two anal spines, along with the attached muscles, venom glands, and integumentary sheath. For medical advice see page 118.

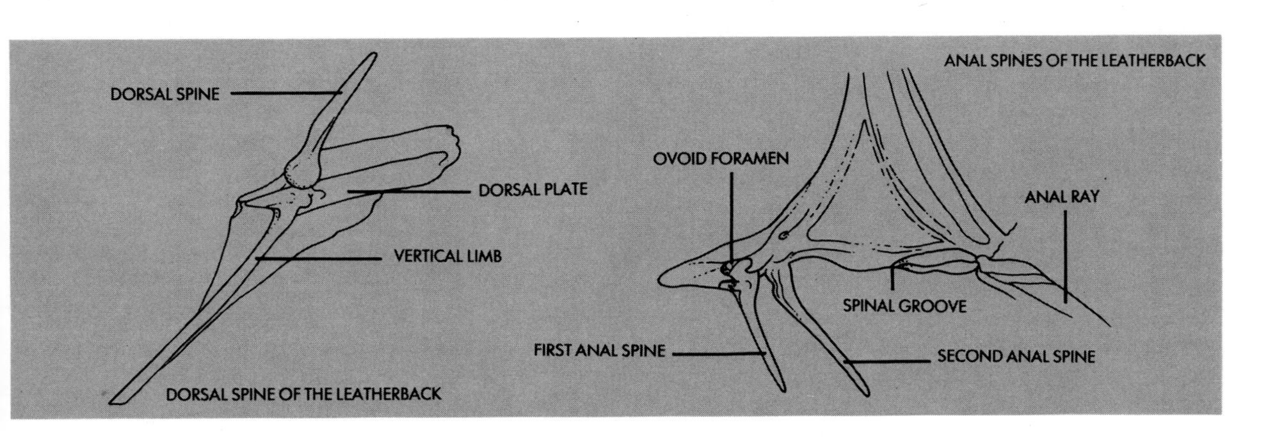

TREATMENT OF VENOMOUS FISH STINGS

To maximize success, treatment should be initiated as rapidly as possible. Efforts in treating venomous fish stings are directed toward combating the effects of the venom, alleviating pain, and preventing secondary infection. Pain follows the trauma produced by the fishes' spines, venom-mediated tissue effects, and the introduction of slime and other irritating foreign substances into the wound. In the case of stingray-induced punctures, the retrorse barbs of the spines may produce severe lacerations with considerable trauma to the soft tissues that may extend into muscle compartments, the thorax and peritoneal cavity. Severe wounds of the extremities should be promptly irrigated with sterile saline or water, if available. Clean seawater may be used as a last resort.

Fish stings of the puncture wound variety, such as those of a catfish or weever, are usually small in size. Rapid removal of the poison is therefore more difficult. It may be necessary to surgically trim the wound edges in order to apply immediate effective irrigation.

Opinion is divided as to the advisability and efficacy of using a ligature in the treatment of fish stings. Arterial occlusive tourniquets are extremely hazardous and should be used only in instances involving a threat to life, when the rescuer is willing to sacrifice a limb in order to save a life. If used, the constriction bandage should be placed at once between the site of the sting (if it is on an extremity) and the trunk, as near to the wound as possible. The bandage should be venous-lymphatic occlusive only, and should be released every 10 minutes for 1 to 2 minutes in order to maintain adequate circulation.

It is recommended that the wound be soaked in non-scalding hot water for 30 to 90 minutes, or until pain is relieved. The water should be maintained at as high a temperature (45°C or 113°F) as the patient can tolerate without skin injury, and the treatment should be instituted as soon as possible. If the wound is on the face or body, hot moist compresses or irrigation with heated fluid can be employed. For further medical advice see Vertebrate Fish Stings, Chapter 7.

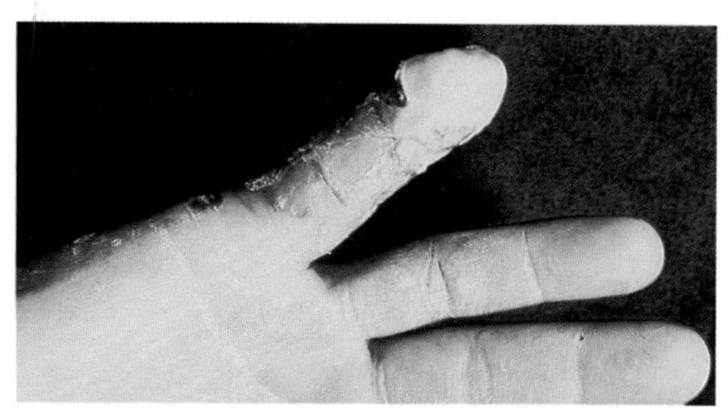

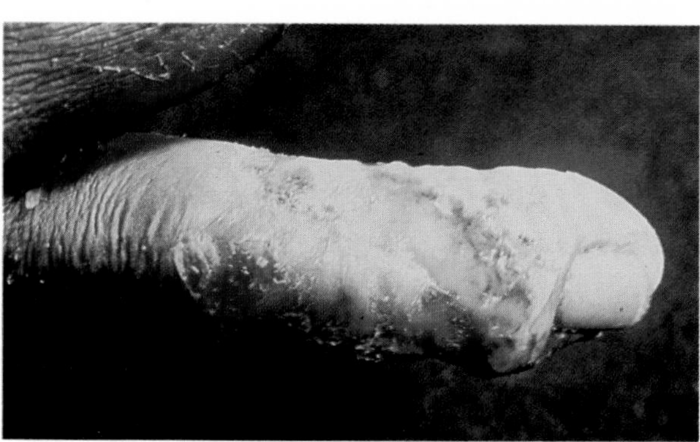

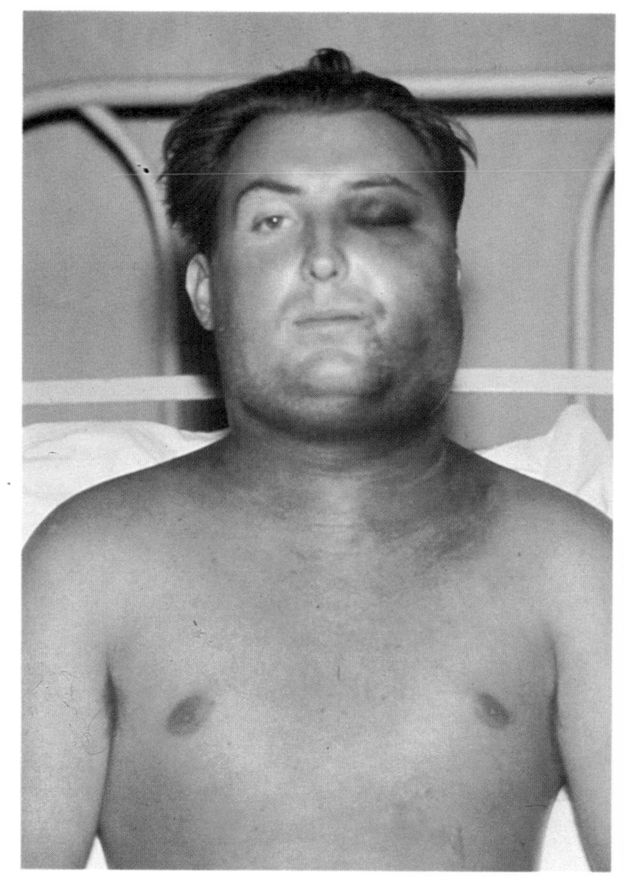

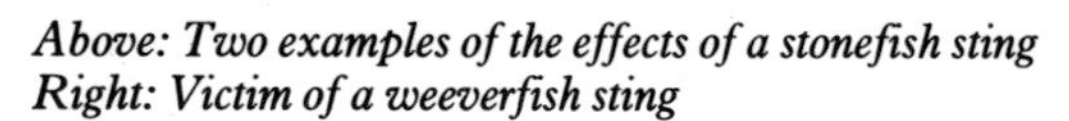

Above: Two examples of the effects of a stonefish sting
Right: Victim of a weeverfish sting

SEA SNAKES

Class REPTILIA
Family HYDROPHIIDAE

The 50 different species of sea snakes are all venomous and capable of inflicting fatal bites if disturbed. Most live in tropical seas, but one is found in fresh water. Most bites are recorded from fishermen removing fish from nets, where the snakes are inadvertently trapped. Occasionally, swimmers are bitten in the muddy water of estuaries or around river mouths where sea snakes are commonly found.

Most sea snakes are about 1 m (39 inches) long, but some are recorded at nearly 3 m (10 feet). With their scaly compressed bodies and paddle-shaped tails, they are well adapted to life at sea and can swim rapidly and efficiently. However, for long periods they lie motionless among the corals and submerged reefs or drift in mid-water. They are capable of moving rapidly either backward or forward with ease, but when placed on land become very awkward, moving ahead with great difficulty.

Sea snakes are commonly found in sheltered coastal waters, especially near river mouths, which seem to be their favorite locations. Around shore areas, these shy but dangerous reptiles may find a rock, crevice, tree root, coral boulder or a piling to hide behind while waiting for prey, which consists mainly of small fish. Sea snakes capture their food near the ocean bottom, usually swallowing prey head-first.

In regions where sea snakes abound, fishermen have been known to catch more than 100 snakes at one time in nets. Although generally found in coastal areas, these animals have been seen 160 to 240 km (100 to 150 miles) from land.

Aipysurus laevis **Lacépède**
Olive-brown sea snake

DESCRIPTION Length 1.8 m (6 feet); tan to light brown or olive-brown body coloration. One of the largest and most heavy-bodied of sea snakes. The body is flattened vertically and may be as thick as a man's arm. The genus is characterized by smooth, overlapping scales.

HABITAT Coastal tropical Australia and New Guinea.

OTHER POINTS Generally docile, but venomous.

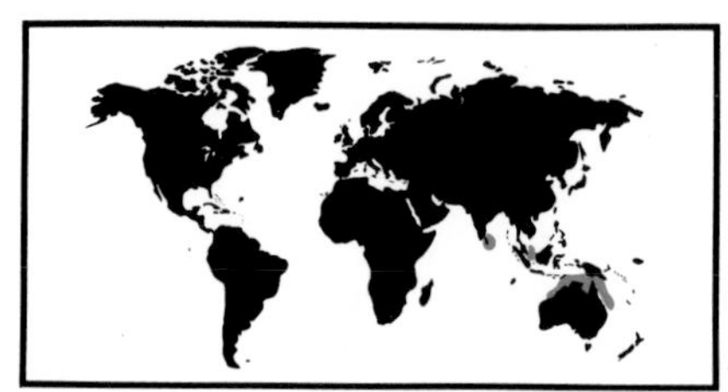

***Astrotia stokesi* (Gray)**
Stoke's sea snake

DESCRIPTION Length 1.6 m (5.2 feet); body coloration is yellowish or light brown, with broad black bands on the upper side.

HABITAT Located along the Makran coast (Pakistan), in Sri Lanka, Singapore, and the north coast of Australia.

OTHER POINTS A mass of Stoke's sea snakes about 3 m (10 feet) wide and fully 95 km (60 miles) long was encountered between the Malay Peninsula and Sumatra. The group was estimated to consist of millions of individuals, and was thought to be either migrating or breeding. The species is venomous.

***Hydrophis cyanocinctus* Daudin**
Sea snake

DESCRIPTION Length up to 1 m (39 inches); sometimes referred to as the blue-banded sea snake due to alternating light to dark blue bands running along the full length of the body.

HABITAT Coastal waters, especially where large rivers flow into the sea; throughout the waters of Indonesia, Japan, the Persian Gulf.

OTHER POINTS This species has caused human fatalities.

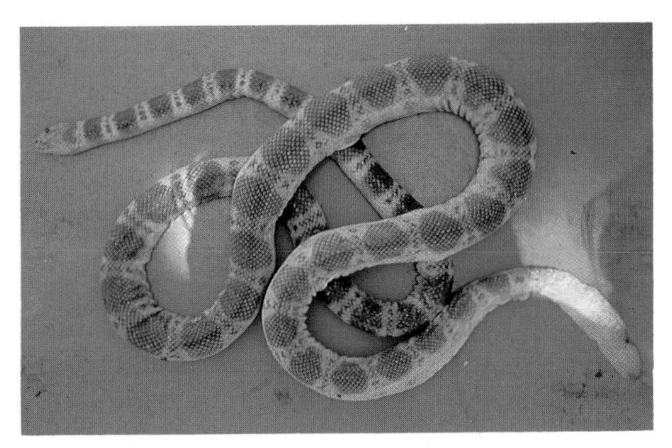

***Hydrophis elegans* (Gray)**
Sea snake

DESCRIPTION Length 2 m (6.5 feet); brown and dull white alternating bands along the full length of the body, with the anterior portion much smaller in diameter than the posterior.

HABITAT Found in tropical Australia.

OTHER POINTS Reputedly harmless, because of its weak striking movement or because of its docility. Nevertheless, the species is venomous, although it has not been incriminated in human envenomation.

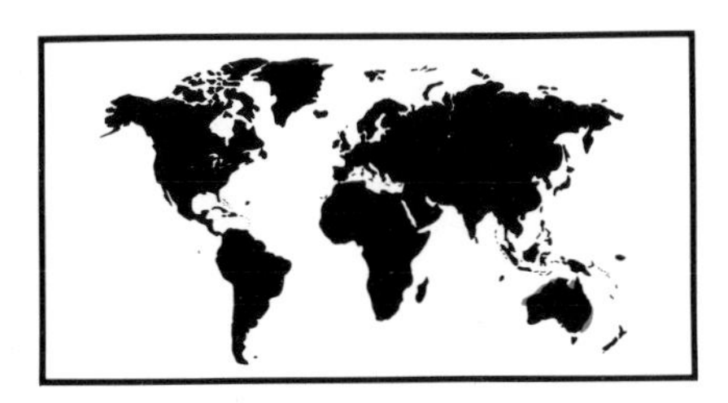

***Lapemis hardwicki* Gray**
Hardwick's sea snake

DESCRIPTION Length 85 cm (33 inches); alternating black and dull white bands. The males have a rough body surface when compared to the smooth scales of the female.

HABITAT From the coasts of southern Japan to the Mergui Archipelago (Burma) and the coast of northern Australia.

OTHER POINTS Though it carries a deadly venom, it has reportedly a somewhat docile nature, with a weak movement. It has a clearly indicated sexual dimorphism.

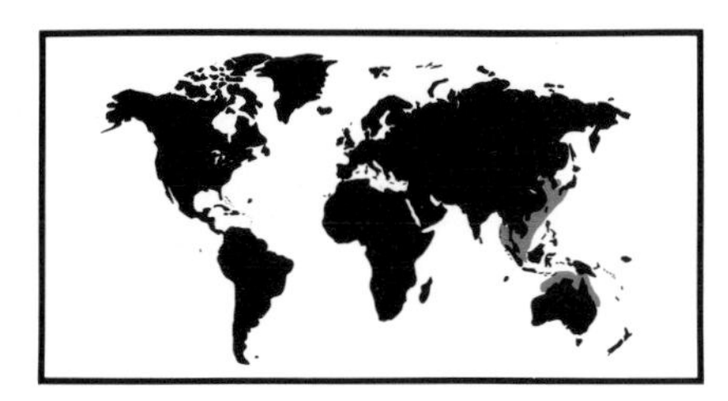

***Laticauda colubrina* (Schneider)**
Sea snake

DESCRIPTION Length 1.4 m (4.5 feet); has alternate black and grayish silver bands along the length of its body.

HABITAT Has been found coiled up in small depressions, under corals or around plant growth and widely distributed throughout the tropical Indo-Pacific region.

OTHER POINTS Extremely venomous but generally it is docile.

***Laticauda semifasciata* (Reinwardt)**
Sea snake

DESCRIPTION Length 1.1 m (43 inches); varies in color, younger specimens having black and bright blue alternating bands, more mature specimens having an almost uniform dull yellow-brown coloration. Older animals seem to lose their stripes.

HABITAT The Philippine Islands, the Moluccas, and the Ryukyu Islands.

OTHER POINTS Although venomous, it has not been positively implicated in human envenomation.

***Pelamis platurus* (Linnaeus)**
Yellow-bellied sea snake

DESCRIPTION Length 85 cm (33 inches); coloration varies, but is typically black on the upper portion of the body and yellow on the underside, with spots on the tail section. Has an elongated snout and a laterally compressed body along its entire length.

HABITAT Has the greatest range of any sea snake, from the west coast of Central America westward to the east coast of Africa, and from Tasmania to southern Siberia. May be found several hundred miles out in the open sea.

OTHER POINTS Despite the small size of its fangs, it is capable of inflicting a fatal envenomation in humans.

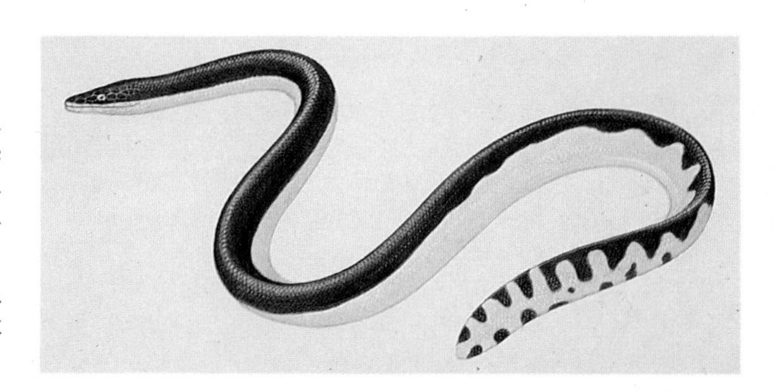

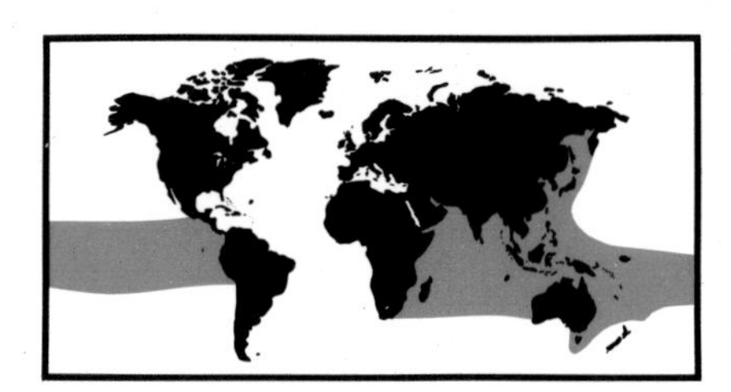

SEA SNAKE BITES

Venom apparatus of sea snakes

Sea snakes have a well-developed venom apparatus which consists of the venom glands and fangs. The venom glands are situated one on either side, behind and below each eye. Most sea snakes have two fangs (of the cobra type) on each side, but some have only one. One drop of sea snake venom has enough poison to kill three adult men. Some species of sea snake can inject up to eight drops of this deadly venom with a single bite.

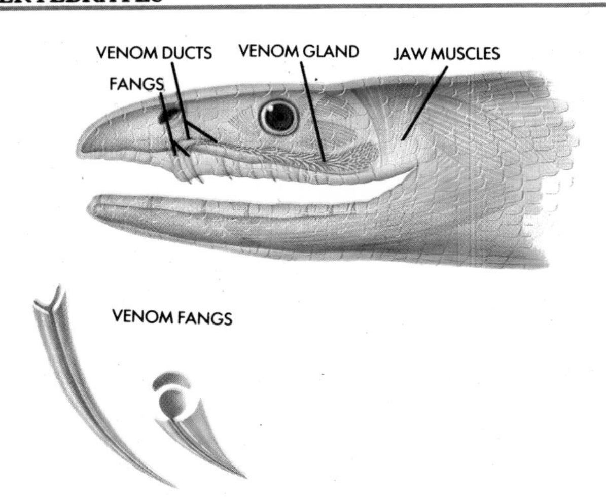

Above: Semi-diagrammatic drawing of the venom apparatus of the sea snake
*Left: **Aipysurus laevis**, the olive-brown sea snake*

Prevention

Sea snakes may occasionally bite bathers. It has been estimated that one sea snake bite occurs per 270,000 man bathing hours in an endemic area such as Penang, Malaysia. The most dangerous areas in which to swim are river mouths, where the sea snakes are more numerous and the water more turbid. The turbidity of the water and resulting poor visibility for the sea snake may contribute to accidental encounters and increased risk of being bitten.

When wading in an area inhabited by sea snakes, shuffling the feet through the bottom sediment to prevent stepping on snakes and to encourage them to swim away can be helpful. It is advisable to remove snakes from fishing nets with a hooked stick or wire. Sea snakes are occasionally captured while fishing with a hook and line. No attempt should be made to remove the hook from the mouth of the snake. The line should be cut and the snake allowed to drop into the water.

Treatment

If treatment is to be effective, it must begin immediately following the bite. First aid and emergency treatment should generally be directed toward retarding absorption of the venom, neutralizing the venom, mitigating the effects produced by the venom, and preventing complications, including secondary infections.

Absorption of sea snake venom is rapid; in many instances it occurs before first aid can be administered. Incision and suction is of value only if it can be applied within the first few minutes following the bite, and is generally not recommended. Therapy for bites from snakes of the family Hydrophiidae is similar to that for terrestrial snakes of the family Elapidae. The affected limb should be immediately immobilized and maintained in a dependent (below the level of the heart) position, while the victim lies down and avoids any form of exertion.

The pressure-immobilization technique for venom sequestration should be employed. If practicable by virtue of the location of the bite, a cloth or gauze pad of approximate dimensions 6-8 cm x 6-8 cm x 2-3 cm (thickness) should be placed directly over the puncture marks and held firmly in place by a circumferential bandage 15 to 18 cm wide applied at lymphatic-venous occlusive pressure (70 mmHg). The arterial circulation should not be occluded, as determined by the presence of distal arterial pulsations and proper capillary refill. The bandage should be kept in place until the victim is brought to a clinic. If the bite is on a digit where a compression bandage cannot be applied and the victim is more than two hours from definitive care, a constriction band which constricts only the superficial venous and lymphatic flow may be applied adjacent to the wound. This should be released for 90 seconds every 10 minutes and should be completely removed after four hours. If the bite is older than 30 minutes, neither technique may be very effective. Ice should not be applied directly to the wound. For further medical advice see Chapter 7.

CHAPTER 4

INVERTEBRATES THAT ARE POISONOUS TO EAT

The sea has always attracted those who seek adventure and never more than at the present time, when skin divers search beneath the waves for the abundant sea life. However, life beneath the sea can be as dangerous as it is fascinating and nowhere more so than in the warm and clear waters of the tropics. Yet with knowledge, many of those dangers can be avoided. Although the sea is a phenomenal source of food, it is not safe to eat everything that may be caught. Obviously, most sea foods can be eaten safely, are nourishing, and form the basis of large industries, but there are some marine animals that contain poisonous chemical substances within their flesh and which may cause illness or even death to a consumer. Unfortunately, the problem is not confined to easily defined groups or even one time of year. Some show seasonal variability in the extent to which they pose a threat. Others become poisonous only after themselves consuming or being exposed to other poisonous substances or organisms. The safest procedure is to become familiar with and adhere to local customs which have evolved in relation to experience. However, even these may not prove infallible.

Many marine organisms may cause food poisoning. Only the major groups are examined here. For convenience of discussion, poisonous marine animals have been divided into those without back bones, and fish and marine mammals (the subject of Chapter 5).

MOLLUSKS

Throughout the world, mollusks are eaten in large quantities, especially the bivalve mollusks which are considered a gourmet's delight. From the oysters of antiquity to the clam chowders of today, huge numbers are regularly eaten with perfect safety. Yet, the problems currently encountered are in part due to the feeding habits of the bivalves as they filter small particles from the water through extensive gills and then concentrate them in the body. Bivalves which thrive in polluted estuarine waters accumulate all manner of pollutants. When our rivers were clean, the problems were slight, but the proliferation of industrial and agricultural wastes has grossly contaminated the bodies of estuarine organisms. This serves to emphasize how careful one must be about the source of sea foods.

Family CARDIIDAE

The cardiids are shallow-dwelling cockles, usually living buried in sand or mud in almost every ocean throughout the world. There are about 200 species in the family.

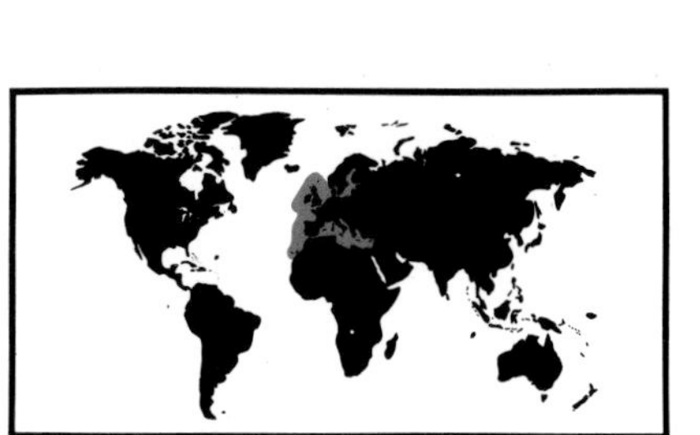

Cardium (Cerastoderma) edule **Linnaeus**
Common cockle

Found in European seas; particularly common in estuarine sands.

Family DONACIDAE

The wedge shells are very widely distributed. There are three genera and about 50 species together making up a family of predominantly suspension filter-feeding bivalves.

Donax serra **(Chemnitz)**
White mussel

Found in South Africa.

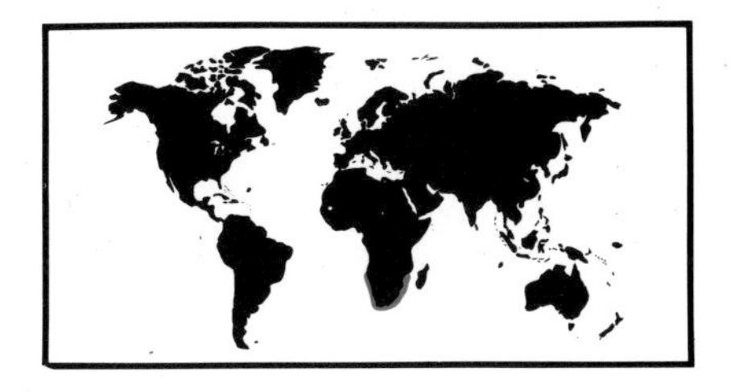

Family MACTRIDEA

The mactrids are surf clams which are shallow-burrowing, filter-feeding bivalves with about 150 species.

Schizothaerus nuttalli **(Conrad)**
Gaper or summer clam

Found from Prince William Sound, Alaska, south to Scammons Lagoon, Baja California, and northern Japan.

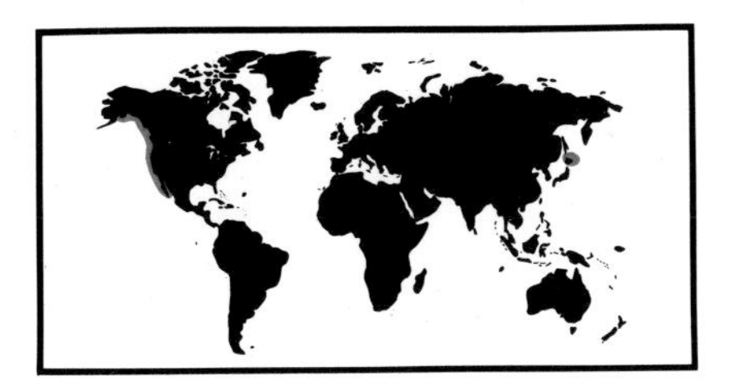

Spisula solidissima **(Darwin)**
Solid surf clam

Occurs from Labrador to North Carolina.

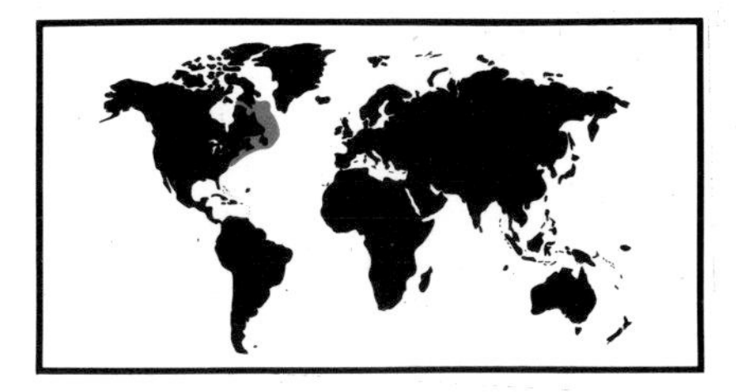

Family MYACIDEA

A family of marine bivalve mollusks comprising the soft-shell clams, with approximately 150 species which are filter feeders. They are of local commercial importance.

Mya arenaria **Linnaeus**
Soft-shelled clam

Found in Britain, Scandinavia, Greenland, the Atlantic coast of North America, south to the Carolinas; Alaska, south to Japan and Vancouver, British Columbia; and the California and Oregon coasts.

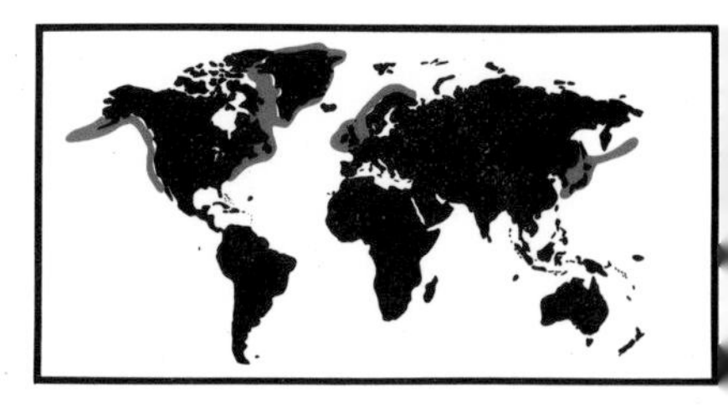

Family MYTILIDAE

This family of marine bivalve mollusks has an elongated shell and lives attached to solid objects, chiefly in the intertidal zone. This family is widely cultured and of commercial importance.

Modiolus modiolus **(Linnaeus)**
Northern horse mussel

Found along the Pacific coast of America from the Arctic Ocean to San Ignacio Lagoon, Baja California; located throughout the Northern Hemisphere.

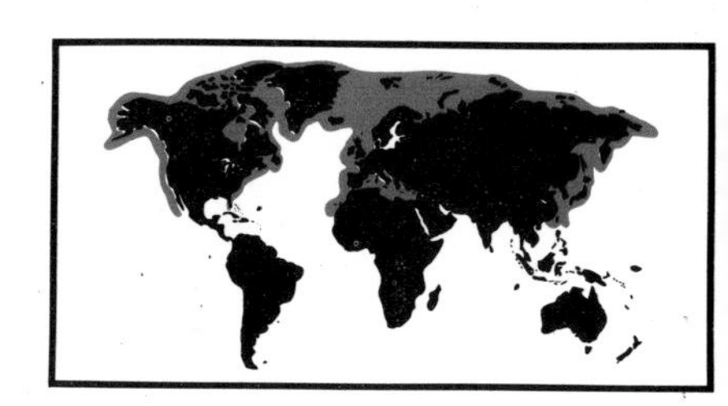

Mytilus californianus **Conrad**
Bay mussel

From Alaska and the Aleutian Islands, eastward and southward to Socorro Island (Chile).

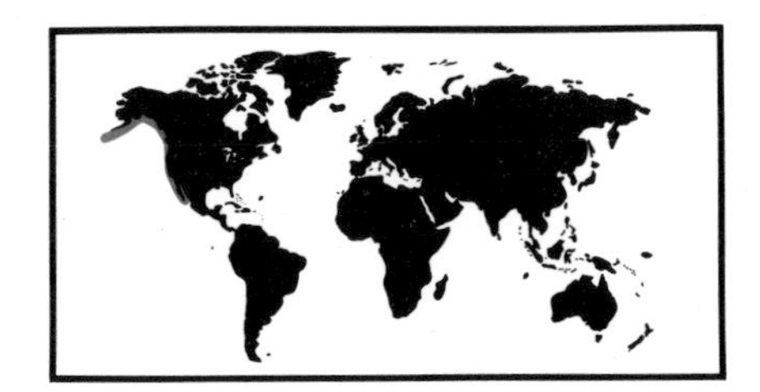

Mytilus edulis **Linnaeus**
Common mussel

Ranges from the Arctic Ocean to South Carolina, Alaska to Cabo San Lucas, Baja California; practically worldwide in temperate waters.

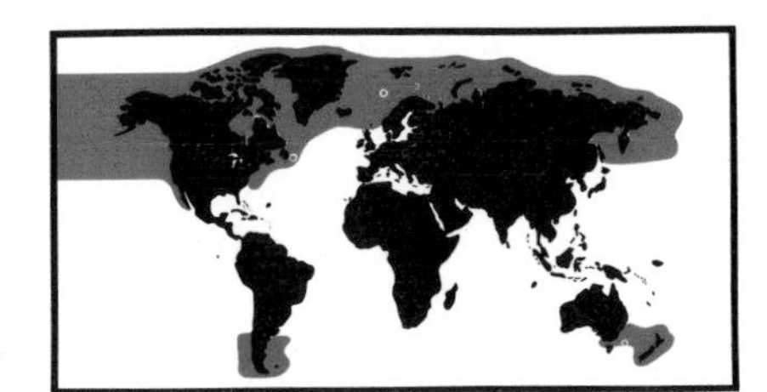

Family SOLENIDAE

This is a family of marine clams with elongated curved shells, comprising the razor clams.

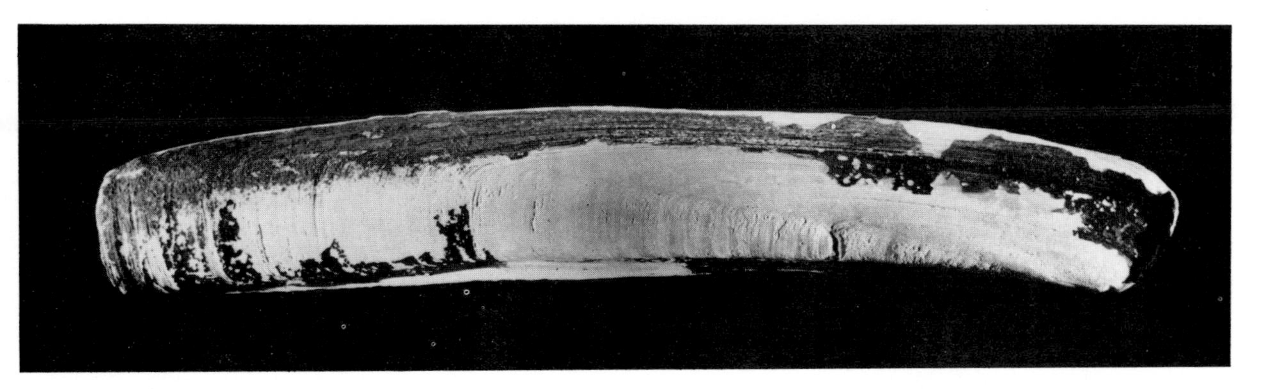

Ensis directus **Conrad**
Atlantic jackknife or razor clam

Ranges from the gulf of the St Lawrence River to Florida.

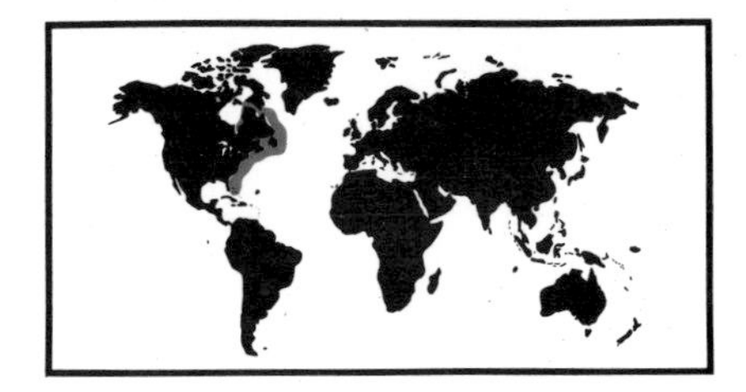

Family VENERIDAE

This is a family of bivalve mollusks mostly having a solid equivalve shell, short siphon, narrow foot, and sometimes a strikingly sculptured shell.

Saxidomus giganteus **(Deshayes)**
Alaska butter clam, smooth Washington, or butter clam

Ranges from Sitka in Alaska to San Francisco Bay, California.

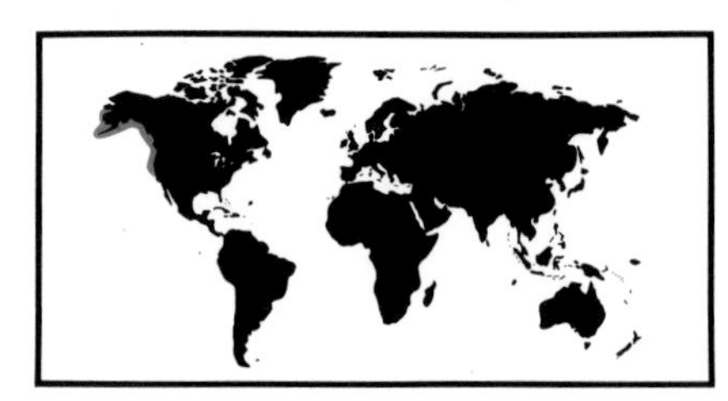

Saxidomus nuttalli **(Conrad)**
Common Washington or butter clam

Ranges from Humboldt Bay, California, to San Quentin Bay, Baja California.

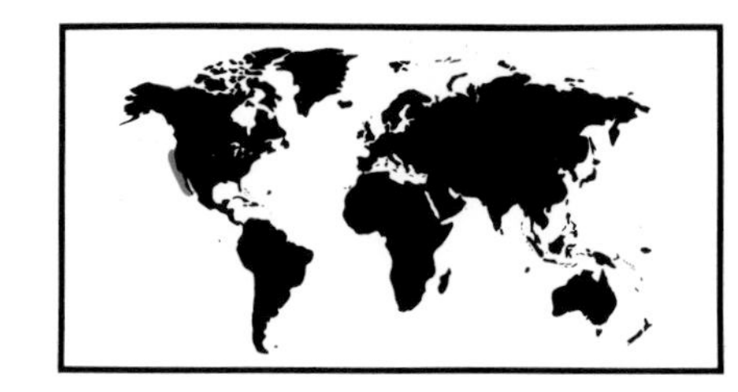

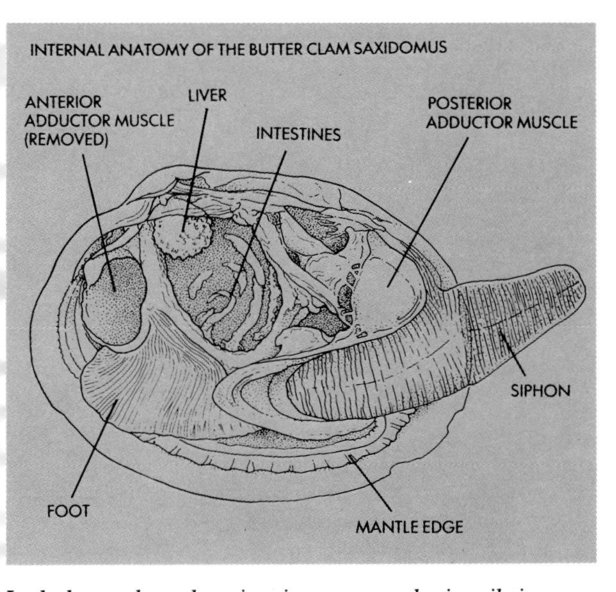

In the butter clam, the poison is concentrated primarily in the siphon

Red tide in Lake Macquarie, New South Wales, Australia

SHELLFISH POISONING

One of the most common shellfish poisonings results from the proliferation of planktonic micro-organisms called dinoflagellates, which may at certain seasons occur in very large numbers in inshore waters. They are filtered from the water by bivalve mollusks, which accumulate the toxins in their tissues. Blooms of dangerous dinoflagellates often result in inshore waters becoming reddish-brown and are referred to as "red tides". The most common of these poisonous dinoflagellates is *Gonyaulax*, which will rapidly multiply in warmer summer months. The Alaska butter clam, *Saxidomus giganteus*, may be poisonous throughout the year. While the poisonous properties of *Gonyaulax* affect many organisms, shellfish are immune to its direct effects and survive to pass the toxin to animals which eat the shellfish.

Human poisoning is of the paralytic type resulting in tingling and burning sensations of the lips, generalized numbness, and eventually difficulty in movement. The paralysis may begin with aching joints and difficulty in swallowing. In serious cases, death can result. Paralytic shellfish poisoning should not be confused with gastrointestinal poisoning, which follows eating shellfish from a contaminated source such as sewage outflow, or from estuaries contaminated with pollutants.

There are many different kinds of dinoflagellate that have been implicated in shellfish poisoning and, as they are difficult to identify, only a few of the commonest species are listed in the table. Although

Dinoflagellate Family	***Species***	***Distribution***	***Toxic Properties***
Gymnodiniidae	*Gymnodinium breve*	North America and Atlantic	Recorded fish kills
Peridiniidae	*Gonyaulax catenella*	North America, Canada, Japan, Chile	Recorded fish kills
	G. tamarensis	North America, Canada, NE Atlantic, North Sea	Recorded fish kills
	G. phoneus	Netherlands	Recorded fish kills
	Pyrodinium bahamense	Brunei, Sabah	Recorded fish kills
Heteraulacaceae	*Gambierdiscus toxicus*	Circumtropical	Implicated in ciguatera fish poisoning

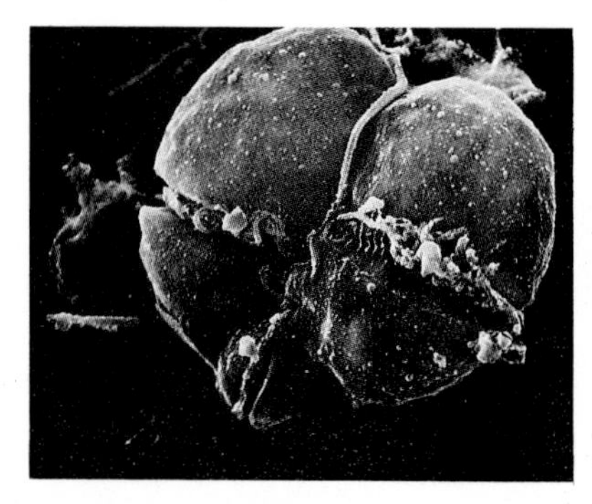

Gymnodinium breve

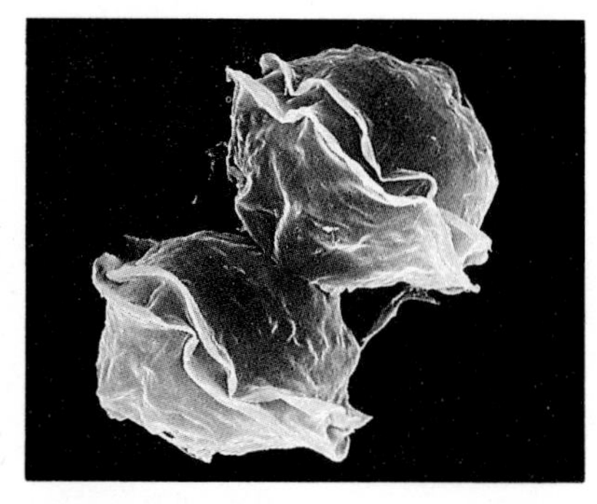

Gonyaulax catenella

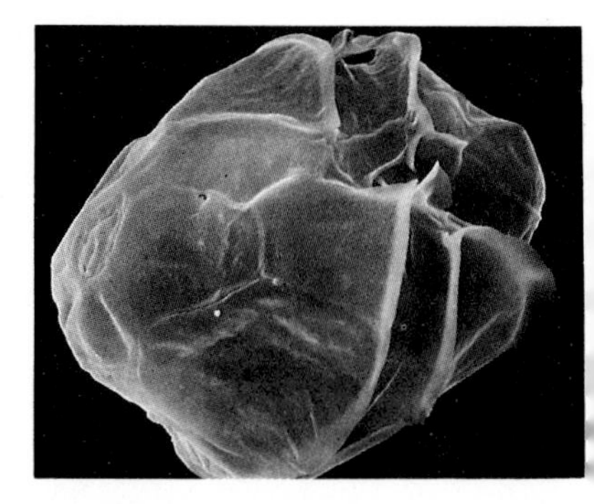

Gonyaulax tamarensis

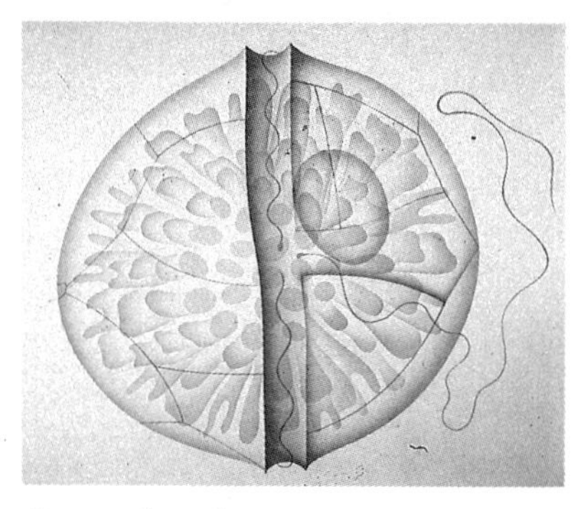

Gonyaulax phoneus

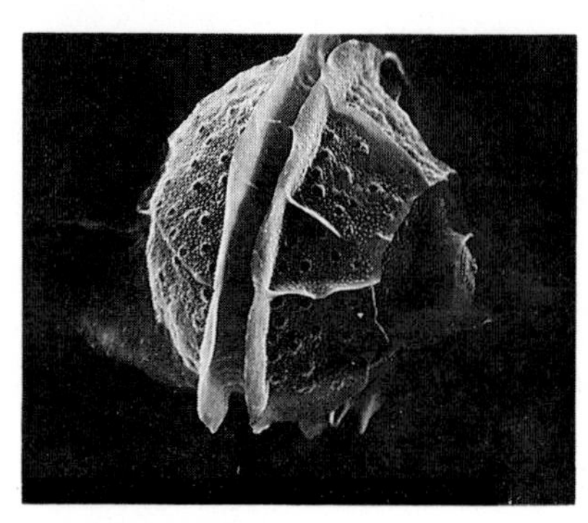

Pyrodinium bahamense

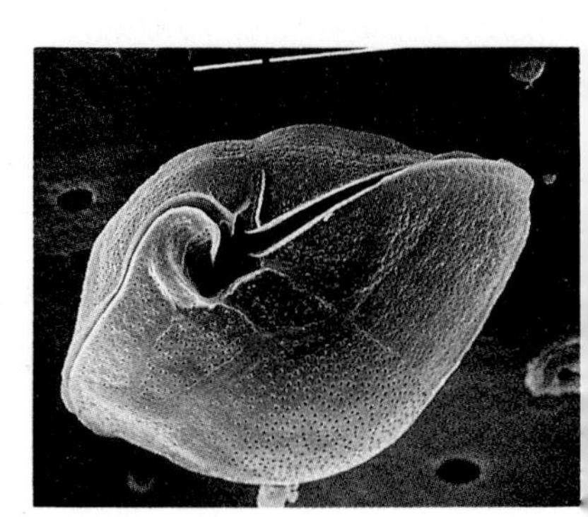

Gambierdiscus toxicus

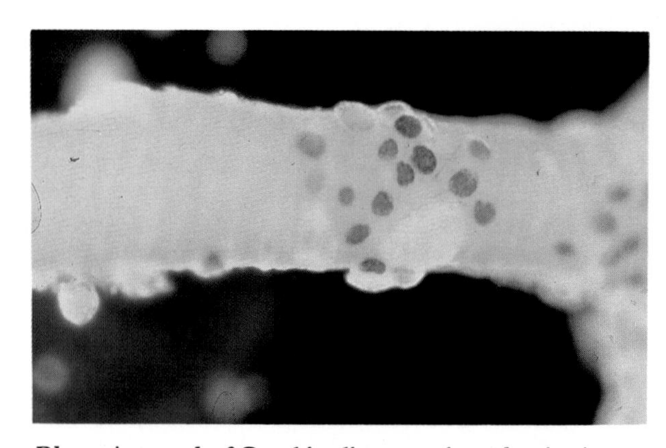

Photomicrograph of ***Gambierdiscus toxicus*** *showing its attachment to the red alga* ***Jania*** *sp*

any bivalve mollusk is potentially capable of accumulating toxic substances, only those living inshore or in estuaries are likely to be extensively used as food sources.

Prevention

The extremely toxic nature of paralytic shellfish cannot be over-emphasized. When toxic shellfish are discovered by local public health authorities, the area is often placed under quarantine. Poisonous shellfish cannot be detected by appearance, odor or taste; only technical scientific procedures can detect the toxins with any degree of certainty.

The digestive organs ("dark meat", gills, and in some shellfish species, siphon) contain the greatest concentrations of the poison. The musculature or white meat is generally harmless; however, it should be soaked or carefully washed before cooking. No part of an animal taken from a quarantined area should be eaten. The broth, or bouillon, in which the shellfish is boiled is especially dangerous, since the poison is water soluble. The broth should always be discarded. The tidal level from which shellfish are gathered cannot be used as a criterion for whether or not shellfish are safe to eat. If in doubt, throw them out.

Treatment

Therapy is supportive and based on symptoms, as there is no specific antidote. Any person who is suspected of having ingested a mollusk contaminated with toxins that cause paralysis should be rushed to a hospital. For further medical advice see Paralytic Shellfish Poisoning in Chapter 7.

COELENTERATES

Coelenterates are generally not very appetizing and it is wise not to attempt to eat them, although jellyfishes are considered a delicacy in the Orient. The stinging cells of most anemones can produce extremely painful lesions if touched by sensitive parts of the body such as the mouth. Despite the obvious danger of attempting to eat sea anemones, case records are scattered throughout the Indo-Pacific region. In Samoa anemones are cooked before being eaten, which reduces the toxicity of some of the poisons. Some of the zoanthids, or soft corals of the genus *Palythoa*, contain a poison called palytoxin, which is produced in the body and tissues of the organism rather than in the nematocysts upon the tentacles. The poison is reported to be detoxified by heat and acid, and is likely to be destroyed if the animal is cooked or immersed in gastric juices. There are reports of swimmers experiencing numbness and tingling of the lips when swimming in closed pools containing large numbers of *Palythoa*. However, it is not known if this results from discharged nematocysts in the water or the presence of palytoxin per se.

The following species have often been implicated in human poisonings, and are taken as respresentative only. Many other species from each family are eaten and have caused similar problems.

Family ACTINIIDAE

This family of sea anemones is characterized by simple tentacles, and is found in many different marine environments. There are an estimated 44 genera and about 225 species.

Physobrachia douglasi **(Kent)**
Lumane sea anemone

DESCRIPTION About 5 cm (2 inches) in diameter; the pedal disc is red, while the tentacles are a greenish color.

HABITAT Polynesia, westward to the Indian Ocean.

OTHER POINTS May inflict a mild urticanal rash.

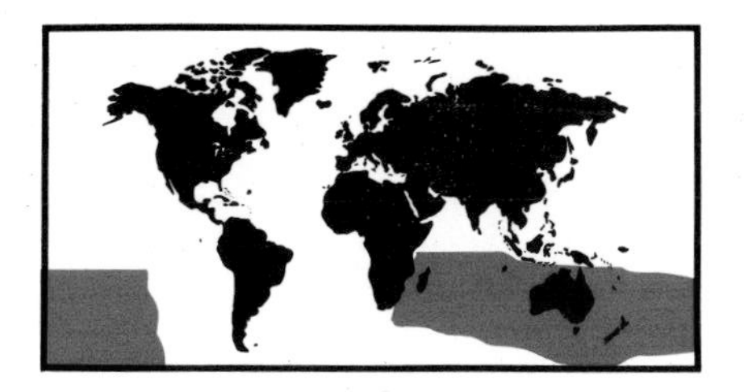

Family ACTINODENDRONIDAE

These actinians have a well-developed pedal disk, which is typically buried in the sand. The family contains three genera with about 10 species.

Rhodactis howesi **(Kent)**
Matalelei sea anemone

DESCRIPTION About 7 cm (2.7 inches) in diameter; varies in shades of purple.

HABITAT Commonly found in coastal areas of Polynesia, westward to the Indian Ocean.

OTHER POINTS Generally considered to be poisonous when raw, but safe to eat when cooked.

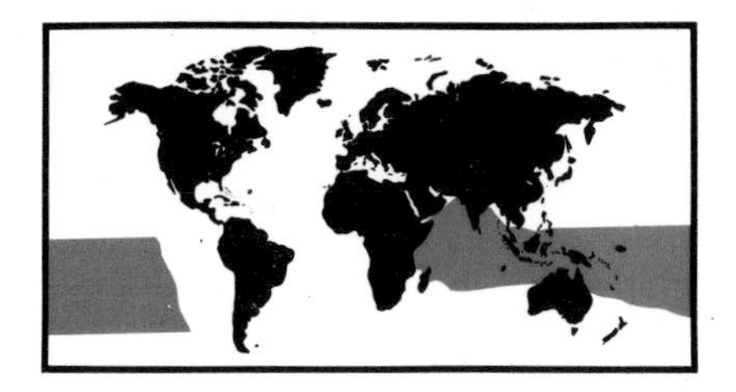

Family STOICHACTIIDAE

This is a family comprised of two genera and approximately 42 species. The tentacles are usually short and wart-like but may be longer depending on the particular genus.

***Radianthus paumotensis* (Dana)**
Matamala samasama sea anemone

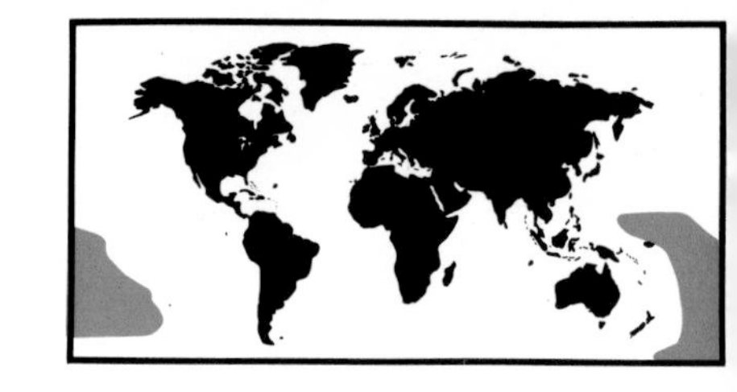

DESCRIPTION Up to 8 cm (3 inches) in diameter; pale coloration with light lavender tips on tentacles.

HABITAT Southern Polynesia, westward to Micronesia.

OTHER POINTS Not dangerous to handle but is reputed to be poisonous if eaten raw.

Family ZOANTHIDAE

Certain species of the zoanthids carry potent toxins. The family is divided into three distinct genera, containing over 150 species.

Palythoa toxica Walsh and Bowers
Deadly seaweed of Hana (Limu-make-O'Hana)

DESCRIPTION Diameter of the disc is 9 mm (0.35 inches); white pedal disc with long and elegant transparent tentacles.

HABITAT This species of soft coral is found in the Caribbean and the tropical Pacific.

OTHER POINTS A poisonous species containing the deadly palytoxin, which has been found to have neurotoxic and cardiotoxic effects in laboratory animals.

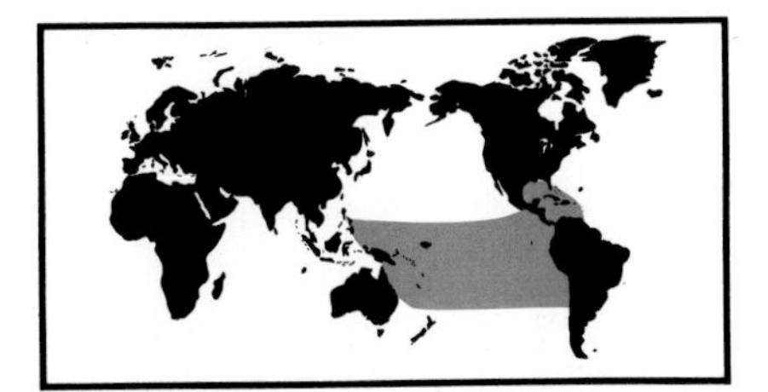

Above left: **Palythoa** *zoanthid anemone*
Above right: **Palythoa tuberculosa**
Left: **Palythoa caesia**

ECHINODERMS

There are only a few recorded cases of echinoderms poisonous to humans, although it is not known whether the symptoms were caused by a toxin manufactured by the echinoderm or by secondary agents such as bacteria. Thus far, only sea urchins and sea cucumbers have been recorded as poisonous to man; most cases of poisoning result from ingestion of sea urchin eggs.

Family ECHINIDAE

There are 31 species in this family of nine genera. *P. lividus* is generally known for its rock-boring habits and as a source of food. Several other species are additionally eaten.

Paracentrotus lividus **Lamarck**
European sea urchin

The diameter reaches 7 cm (2.7 inches). Found along the Atlantic coasts of Europe and the Mediterranean. The eggs are frequently eaten and there are records of some poisonings.

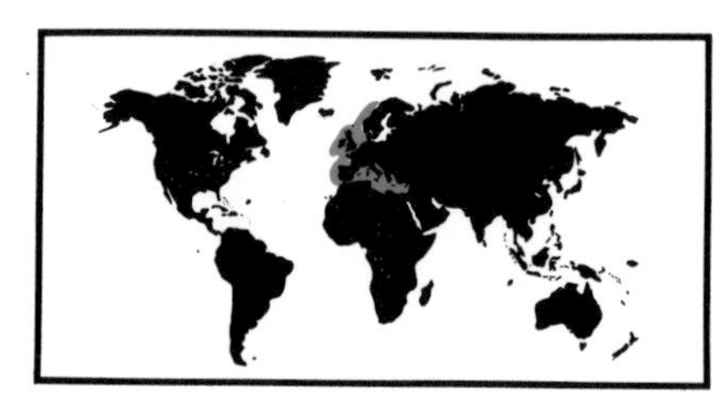

Family TOXOPNEUSTIDAE

Most of the approximately 35 species are in 10 genera and are found in large numbers in shallow coastal waters of the Caribbean Sea.

Tripneustes ventricosus **(Lamarck)**
White sea urchin

It is believed that white sea urchins acquire their poisons from the algae upon which they feed. The diameter reaches 7 cm (2.7 inches). Found in the West Indies.

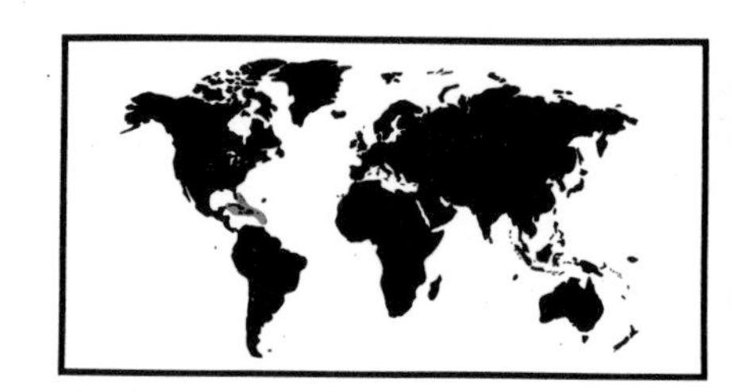

SEA CUCUMBER POISONING

Sea cucumbers are free-living echinoderms, having an elongated body without free arms, but with a series of tentacles encircling the mouth, located at the anterior end of the body. Sea cucumbers, members of the class Holothuroidea, serve as important items of food in some parts of the world, where they are sold under the name of "trepang" or "beche-de-mer". They are boiled and then smoked or dried in the sun. Trepang is used to flavor soups and stews.

When disturbed, sea cucumbers discharge a sticky fluid from large glands of Cuvier. This cantharidin-like substance irritates the skin and eyes of humans. Poisonings and occasional deaths from eating sea cucumbers have been reported.

All of the sea cucumbers illustrated below have been found to contain holothurin, a toxic saponin. Visceral liquid ejected by some sea cucumbers may produce a skin rash or blindness and severe irritation if brought into contact with the eyes. Although certain cucumber species are commonly eaten, ingestion should be considered with caution.

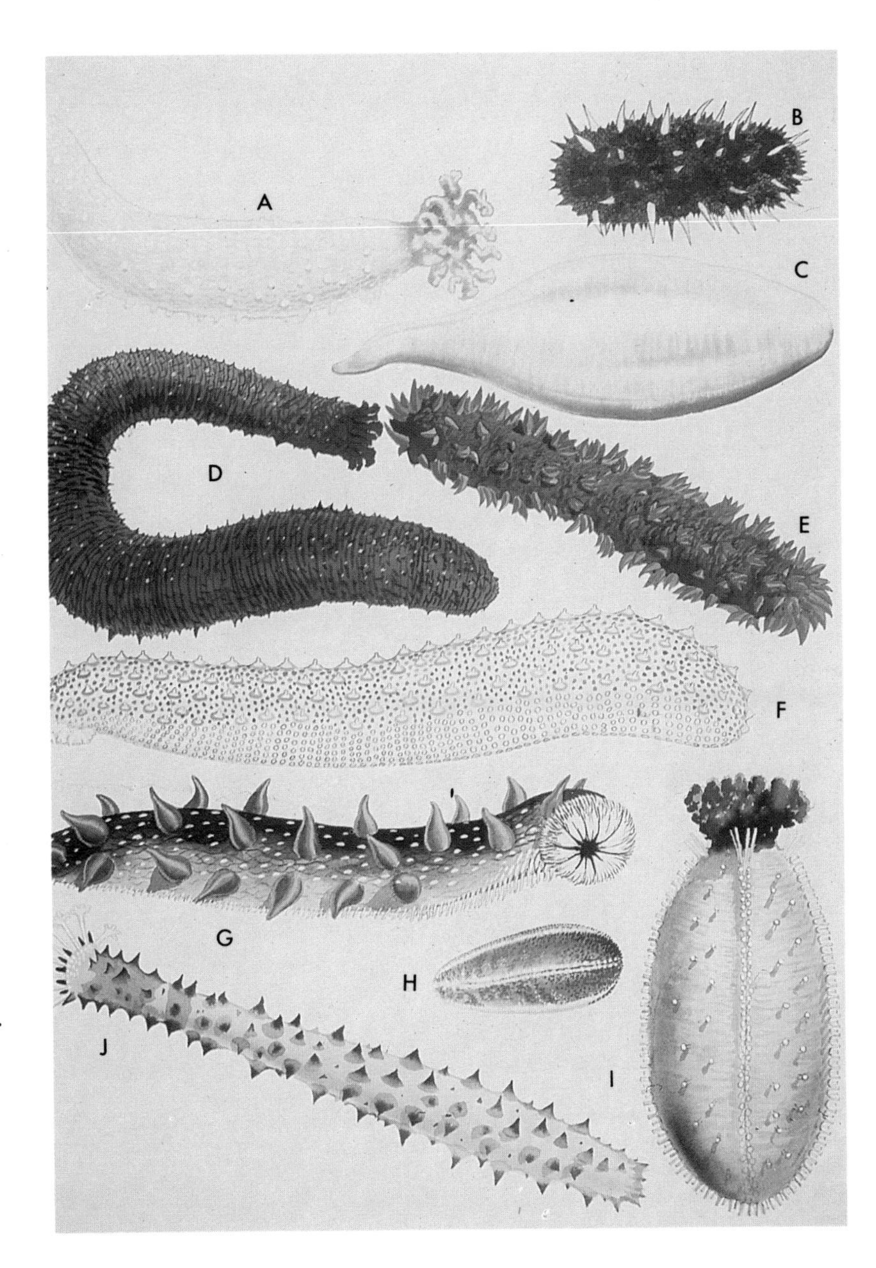

Poisonous sea cucumbers.

a. ***Afrocucumis africana*** **(Semper).** Length 8 cm (3 inches). Japan.

b. ***Stichopus japonicus*** **Selenka.** Length 40 cm (16 inches). Indo-Pacific, Australia.

c. ***Paracaudina chilensis*** *var.* **(Von Marenzeller).** Length 20 cm (8 inches). California, Mexico, Guatemala, westward through the Indo-Pacific, China, Japan and the Aleutian Islands.

d. ***Holothuria vagabunda*** **Selenka.** Length 20 cm (8 inches). Tropical western Pacific.

e. ***Thelenota ananas*** **(Jaeger).** Length 75 cm (29 inches). Japan.

f. ***Holothuria monocaria*** **Lesson.** Length 40 cm (16 inches). Japan.

g. ***Stichopus japonicus*** **Selenka.** Length 30 cm (12 inches). Japan.

h. ***Pentacta australis*** **(Ludwig).** Length 8 cm (3 inches). Japan.

i. ***Cucumaria japonica*** **(Semper).** Length 20 cm (8 inches). Japan.

j. ***Holothuria impatiens*** **(Forskål).** Length 40 cm (16 inches). Australia.

MARINE ARTHROPODS

Phylum ARTHROPODA
Class MEROSTOMATA

The phylum Arthropoda (invertebrate animals with jointed legs) is the largest single group in the animal kingdom, having more than 800,000 species. Relatively little is known about poisonous marine arthropods. However, Asiatic horseshoe crabs and some tropical lobsters and reef crabs may be quite poisonous, probably because secondary toxins have been absorbed by the crustaceans during their own feeding activities.

Family LIMULIDAE

The horseshoe crabs are characterized by an arched cephalothorax, a horseshoe-shaped carapace, and a wide segmented abdomen. They are scavengers which inhabit coastal waters on sandy or muddy bottoms, and range from medium to large-sized species, which usually move by crawling but are capable of a clumsy inverted swimming action. This family consists of three genera with four species.

Carcinoscorpius rotundicauda **(Latreille)**
Asiatic horseshoe crab

Length up to 33 cm (13 inches). Poisoning follows ingestion of unlaid green eggs or the inner parts of this crab during the reproductive season. Most poisonings have occurred in Southeast Asia. Even though they are sometimes poisonous, the large masses of green unlaid eggs are considered a very popular food by many Asiatic people.

Class CRUSTACEA
Family COENOBITIDAE

This family of hermit crabs is represented by 17 species of two genera. They are semi-terrestrial, spending much time on land where they play an important role as scavengers. Like the majority of crustacea, they must return to water to breed. The group is variable in size, ranging from a few centimeters to over 30 cm (12 inches). The largest member of the Coenobitidae is the coconut or robber crab, which, unlike other members of the group, does not use a shell for protection.

Birgus latro (Linnaeus)
Coconut crab

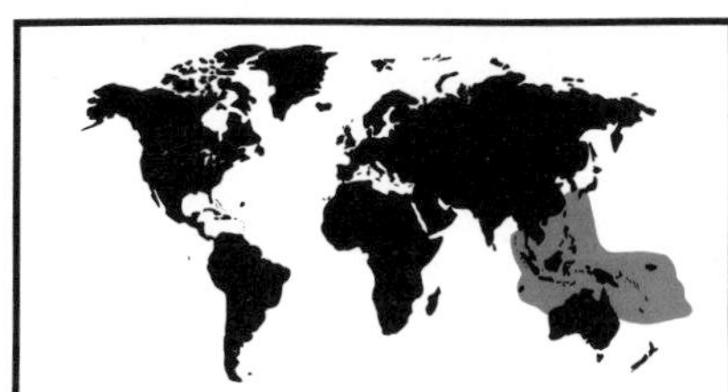

It reaches over 30 cm (12 inches) in length and is largely terrestrial, living throughout the Indo-Pacific region in damp jungle litter and on coral atolls. It feeds almost exclusively on plant material, emerging in the evening to scavenge a variety of plants. It is widely used as a food, though records indicate that the flesh may become poisonous after the crab has eaten certain plants. Poisonings with some fatalities have been reported from the Ryukyu Islands off Taiwan as well as in Tahiti.

Family PALINURIDAE

Members of this family inhabit coral reefs and rocky areas. The family is comprised of eight genera, with about 35 species.

***Panulirus* sp.**
Spiny lobster

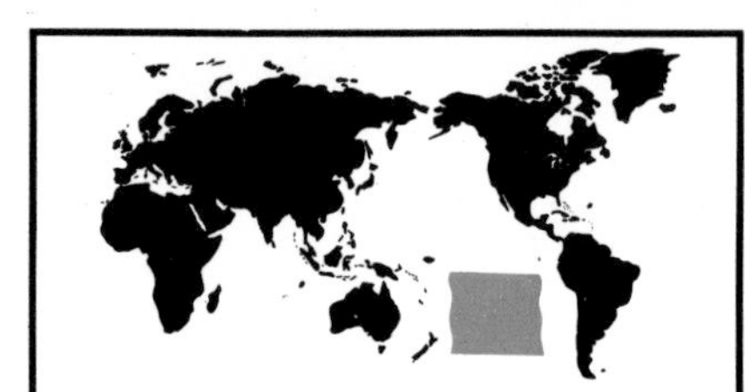

This genus is of considerable commercial importance and several fisheries in different parts of the world depend upon it. Typically an inhabitant of rock and coral reefs, it is found throughout the tropics and in the northeast Atlantic.

Several reports have appeared of poisonous lobsters in southern Polynesia, but little is known about this type of poisoning.

Family XANTHIDAE

The mud crabs are most commonly found on tropical reefs, though their name may imply otherwise. They comprise the largest family of crabs, with more than 100 genera and approximately 1000 species.

Atergatus floridus **(Linnaeus)**
Reef crab

This tropical reef crab is found in the Indo-Pacific and is reported to have caused several poisonings in humans.

Carpilius maculatus **Linnaeus**
Red-spotted crab

It has caused serious poisonings. Found in the Indo-Pacific.

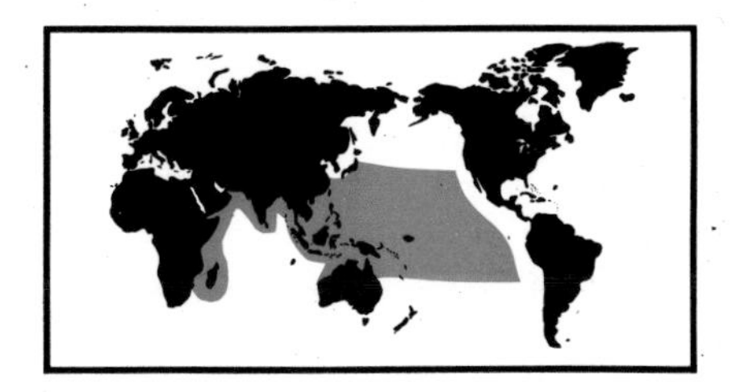

Demania toxica **(Garth)**
Reef crab

This crab has caused violent poisonings in humans. It ranges throughout the Indo-Pacific region.

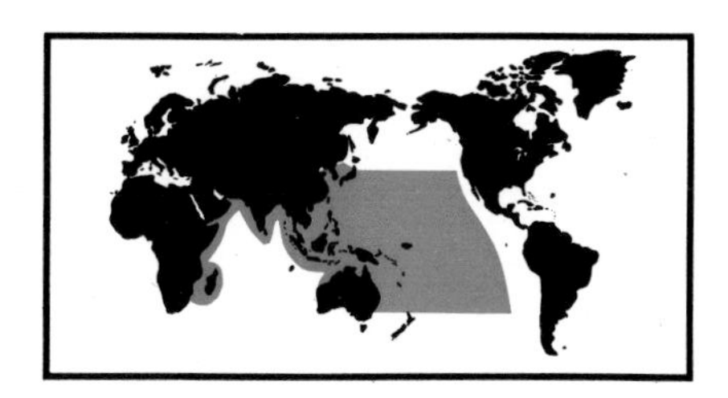

CHAPTER 5

VERTEBRATES THAT ARE POISONOUS TO EAT

While there are many recorded cases of humans suffering gastrointestinal complaints as a result of eating marine fish, in most cases this is traced to secondary contamination of the food. However, there are numerous species in the tropics which are poisonous. The toxins found in some tropical reef fishes may prove fatal.

There are several types of ichthyosarcotoxism, or poisoning resulting from eating fish flesh, that have been designated as follows:

Elasmobranch poisoning – caused by eating sharks, rays, and some of their relatives.

Ciguatera fish poisoning – caused by eating various species of tropical reef fishes.

Scombroid fish poisoning – caused by eating tunas and related species of scombroid fishes (and recently some "non-scombroid" fishes) which have not been properly preserved.

Gempylid fish poisoning – caused by eating castor oil fish of the family Gempylidae.

Hallucinogenic fish poisoning – caused by eating certain tropical reef fishes which produce hallucinations.

Puffer fish poisoning – usually caused by eating the flesh of certain kinds of puffer fishes of the family Tetraodontidae, although other marine organisms have been involved.

Turtle poisoning – caused by eating the flesh of certain types of marine turtles.

Marine mammal poisoning – eating the flesh and various organs of certain marine mammals can also cause poisonings. This includes polar bears, seals, dolphins, porpoises and whales.

ELASMOBRANCH POISONING

This form of poisoning is most commonly caused by eating the livers of sharks. However, the flesh of some large tropical sharks and the Greenland shark, which lives in Arctic waters, has been reported as having caused intoxications in humans as well as in sled dogs.

Family CARCHARHINIDAE

The requiem sharks constitute the largest family of sharks, having at least 13 genera and numerous species. Most carcharhinids are inhabitants of tropical or warm-temperate waters, but may enter cooler zones during the summer months or during part of their annual migrations. Some are circumtropical in distribution, whereas others are confined to narrow distributional areas. Most of the species are harmless, but a few have been incriminated as "man eaters". The requiem sharks are carnivorous, and feed on a great variety of fishes, sharks, stingrays, turtles, birds, sea lions, porpoises, gastropods, horseshoe crabs and crustaceans, and will consume carrion and garbage. Carcharhinids inhabit a variety of different biotopes, ranging from open seas and coastal waters to estuaries, inlets, river mouths and lagoons. One species, *Carcharinus nicaraguensis*, is known only from Lake Nicaragua, a freshwater lake.

On numerous occasions, carcharhinids have been observed pursuing schools of small fishes in shoals where the water was so shallow that the backs of the sharks were almost completely exposed to the air. Their development is either of the viviparous or ovoviviparous type and some species may attain a length in excess of 8 m (26 feet). The liver and flesh of some of these sharks have caused human intoxications.

Carcharhinus melanopterus **(Quoy and Gaimard)**
Blacktip reef shark

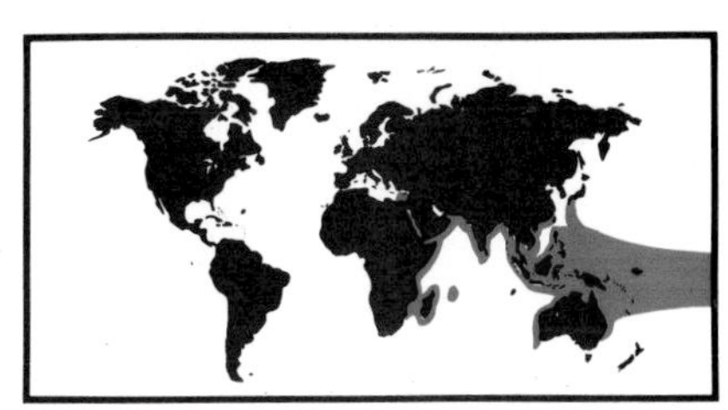

Very abundant and found in shallow lagoons and coastal waters in the Indo-Pacific, South Africa and the East Indies, and the Hawaiian, Tuamotu, and the Marianas Islands.

Family DALATIIDAE

The sleeper sharks constitute one of the small families of sharks, consisting of approximately eight genera with about 11 species. They have a range which extends from the Arctic waters of Greenland to South Africa and the Indo-Pacific.

Somniosus microcephalus **(Bloch and Schneider)**
Greenland shark

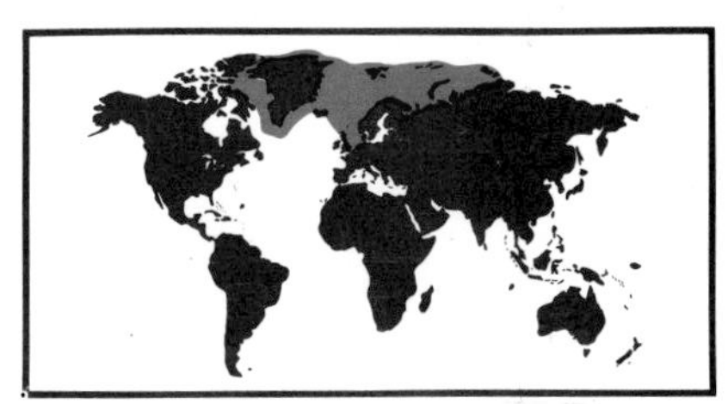

The single incriminated toxic family member, *Somniosus microcephalus*, is an inhabitant of Arctic and north temperate waters. It has a wide bathymetric range, from the surface to 1200 m (660 fathoms) or more. *Somniosus* is considered of commercial value in Norway, Iceland and Greenland, where it is sought for its liver oil. The flesh is used for dogfood and, when dried, is occasionally eaten by humans. The flesh is said to be toxic when fresh.

Family HEXANCHIDAE

This family consists of three genera and four species. Cow sharks are widely distributed representatives of the family occurring throughout most temperate and tropical seas. They are largely bottom-dwellers and somewhat sluggish in movement. Their bathymetric range is considerable. *Hexanchus grisseus* has been taken at more than 1300 m (700 fathoms), whereas other species seem to prefer shallow water.

They are said to be voracious feeders, eating fishes and various crustaceans. *H. grisseus* may reach a length of more than 4 m (13 feet), but *Heptranchias perlo* is reputed to have a maximum length of about half this at 2 m (6.5 feet). Cow sharks are ovoviviparous in development.

Heptranchias perlo **(Bonnaterre)**
Seven-gilled shark

Eastern and western Atlantic, the Mediterranean Sea, the Cape of Good Hope and Japan.

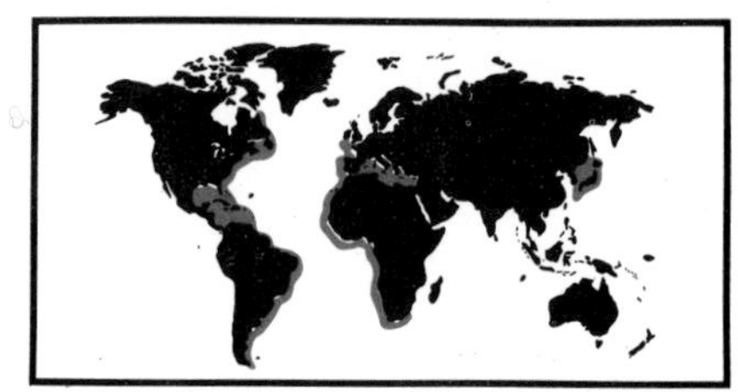

Hexanchus grisseus **(Bonnaterre)**
Six-gilled shark

Atlantic, Pacific coast of North America, Chile, Japan, Australia, southern Indian Ocean, and South Africa.

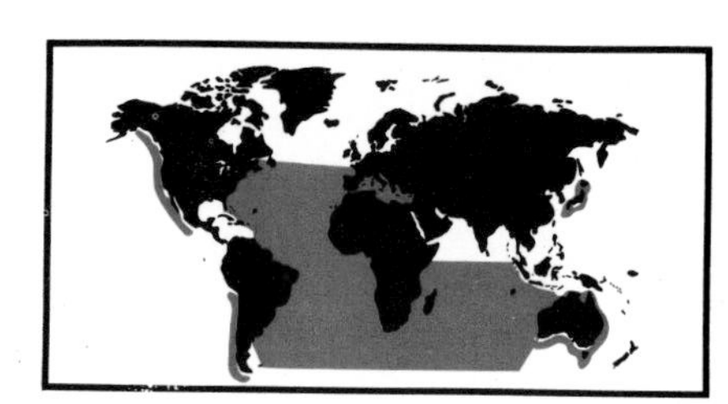

Family ISURIDAE

This family is made up of three genera and about five species. The only mackerel shark which may be toxic is the great white shark *Carcharodon carcharias*, which is cosmopolitan in all warm seas. This species has the reputation of being one of the most voracious of fishlike vertebrates, devouring sharks, large fishes, sea lions, turtles and garbage. The species has been known to charge small boats without provocation and has a reputation as a man eater. Although typically an inhabitant of the open seas, it is not averse to entering shallow water to feed. It attains a length of 5 m (16 feet) or more. Its reproduction is ovoviviparous.

Carcharodon carcharias **(Linnaeus)**
Great white shark

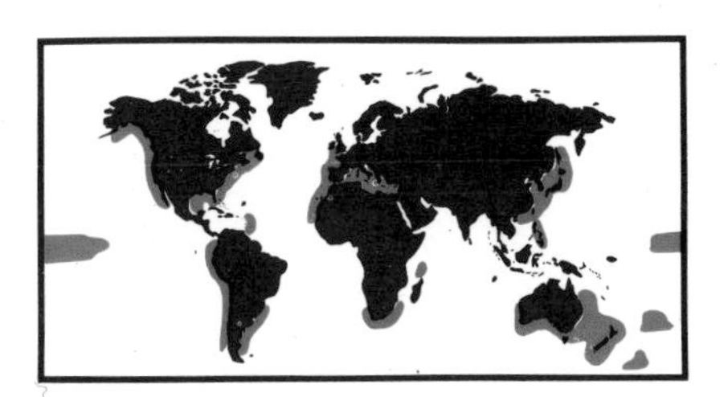

Family SPHYRNIDAE

This family has one genus with nine species. The hammerhead sharks, as their name implies, are characterized by having the anterior portion of the head greatly flattened dorsoventrally and widely expanded laterally in a hammer-like shape. The eyes are situated at the outer edges of the head. Members of the family are found in all warm seas. Hammerhead sharks are known to be man eaters. Members of this group may be found far offshore, swimming near the surface of the water, but not infrequently are encountered inshore, in bays and lagoons. Their diet consists largely of fish, crustaceans and squid. They attain a length of 4 m (13 feet) or more. Some species are viviparous, whereas others are believed to be ovoviviparous.

Sphyrna zygaena **(Linnaeus)**
Smooth hammerhead shark

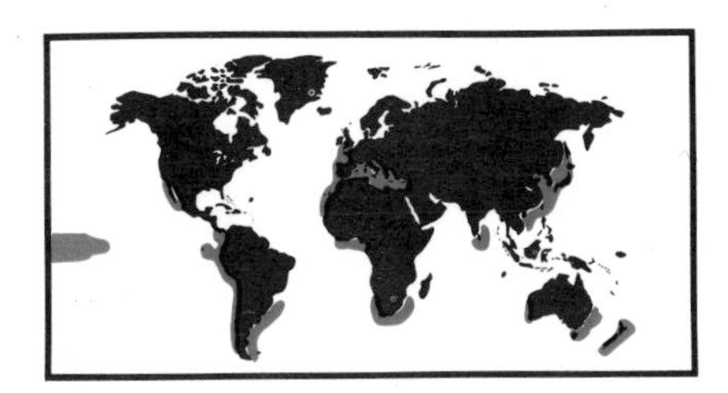

CIGUATERA FISH POISONING

Ciguatera poisoning is one of the most serious and widespread forms of fish poisoning. It originates mainly from tropical inshore fishes, with more than 400 species being implicated. Because many of these fish are regarded as valuable food, those contaminated with ciguatera toxins are a serious threat to local tropical fisheries. Without warning, fish which are normally safe to eat may become poisonous and remain dangerous for years. Ciguatoxin accumulates through the food chain as fish feed on toxic dinoflagellates and algae, invertebrates or fishes. The drawing below illustrates how ciguatoxin occurs in the food chain.

There is now excellent evidence that the dinoflagellate *Gambierdiscus toxicus*, frequently found on the surface of benthic brown seaweed and upon other algae species, may produce ciguatoxin. Dinoflagellates may be often eaten by filter-feeding invertebrates, plankton-feeding fishes, plant-eating fishes and indirectly by fishes eating other fishes. Ciguatera poisoning is most commonly found in subtropical regions, but the greatest number of ciguatoxic fishes are near the islands of the tropical Pacific, Caribbean and Indian Oceans.

The fish presented in this section are generally safe to eat and form the basis of many local fishing industries. However, in certain areas these same species may become poisonous. Accordingly it is often difficult to predict when and where ciguatera poisoning will occur.

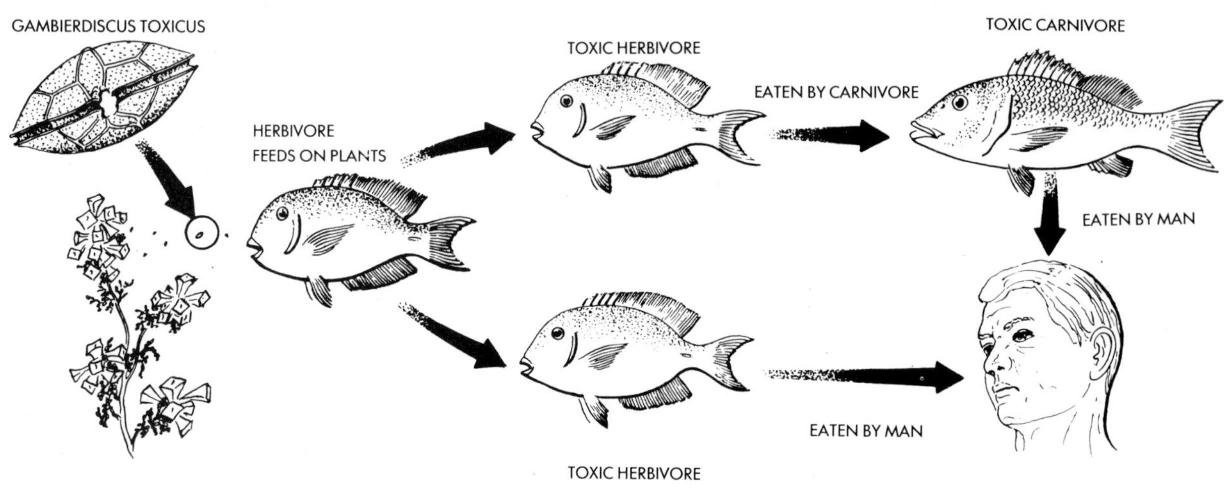

Above: Diagram showing the ways in which ciguatoxin enters the food chain
Right: Geographical distribution of ciguatoxic fishes

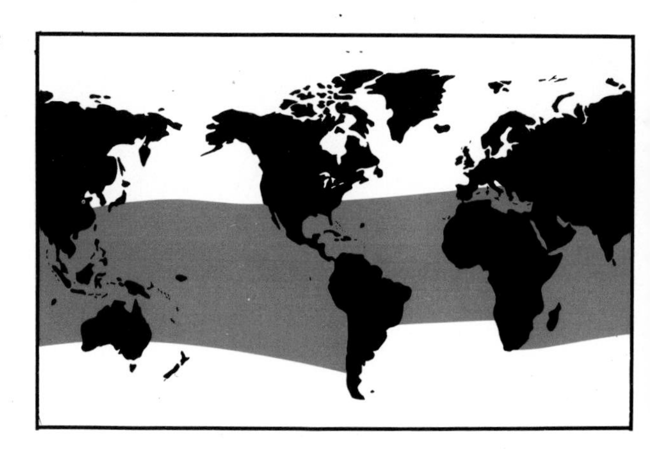

Family ACANTHURIDAE

Surgeonfishes or tangs are a group of shore fishes inhabiting warm seas. The principal genus *Acanthurus* is characterized by the presence of a sharp, lancelike and movable spine which is located on each side of the caudal peduncle. Surgeonfishes are particularly common in surge channels and shoal areas in the vicinity of coral patch reefs. They are predominantly herbivores, feeding on fine filamentous algae of numerous species. Small fishes and invertebrates may be ingested. Most surgeonfishes are small to moderate in size. The eggs and larvae of some species are pelagic. The family is comprised of 10 genera, with about 77 species.

Acanthurus glaucopareius Cuvier
Surgeonfish

Length 20 cm (8 inches). Indonesia, Philippine Islands, and tropical Indo-Pacific.

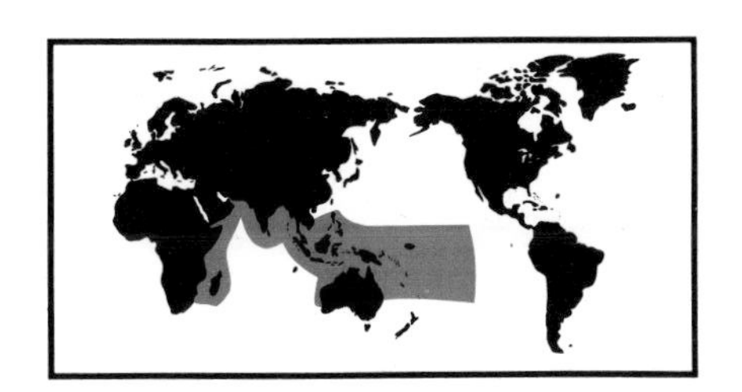

Acanthurus triostegus (Linnaeus)
Surgeonfish

Length 20 cm (8 inches). Indo-Pacific region.

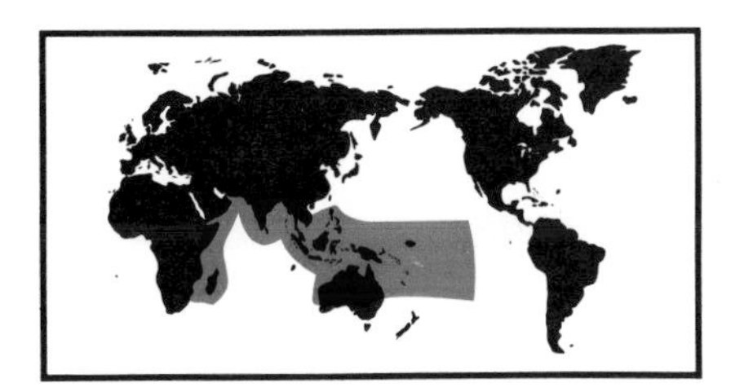

Family ALBULIDAE

The ladyfishes, or bonefishes, are found in most warm seas and are common along sandy coasts. Occasionally they are seen in large schools. They prefer shallow water over low tidal flats. Their food consists of shellfish, crustaceans and worms, which they take from the mud and grind up with their pavement-like teeth. Bonefish are considered a good game fish in some areas, and are eaten to a limited extent. There are two genera with four species.

Albula vulpes **(Linnaeus)**
Ladyfish

Length 1 m (39 inches). Lives in all warm seas.

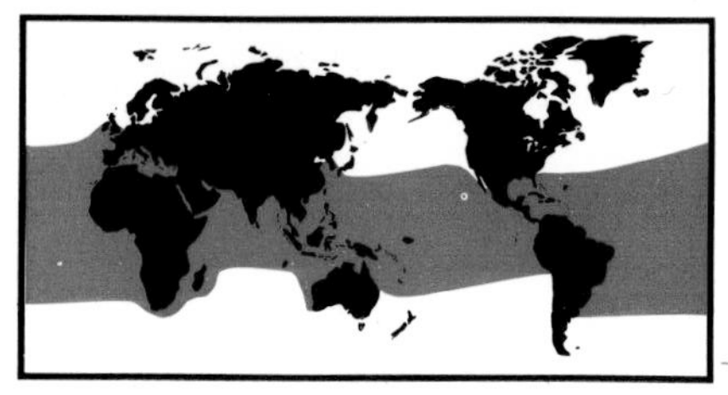

Family ALUTERIDAE

Filefishes are a group of small-sized, shallow-water shore fishes of temperate and tropical seas, characterized by compressed bodies covered by small prickles. The musculature is generally too meager to make these fishes useful as food, and the flesh is considered to be dry, bitter and offensive in taste. However, some natives dry and eat them. The family Aluteridae is sometimes included in the Monacanthidae family. The habits of the fishes of these two families are similar. There are 12 genera, with perhaps 85 species.

Alutera scripta **(Osbeck)**
Filefish

Found in all warm seas.

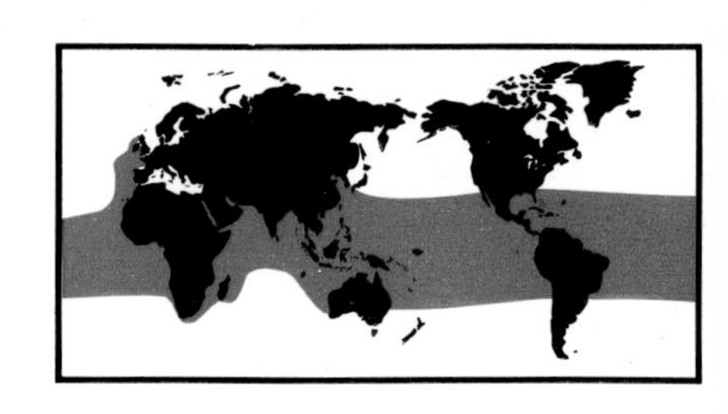

Family BALISTIDAE

Triggerfishes are a group of small to moderate-sized coastal fishes, characterized by a deep, compressed body covered with a thick layer of enlarged bony scales. The first two dorsal spines are modified into a trigger-like device. They are widely distributed throughout all warm seas and prefer shallow reef areas, although some are found in deeper water. They are omnivorous in their eating habits, ingesting corals, sponges, urchins, algae, and various other small organisms. Despite a slow swimming movement, they migrate over long distances by floating with the currents.

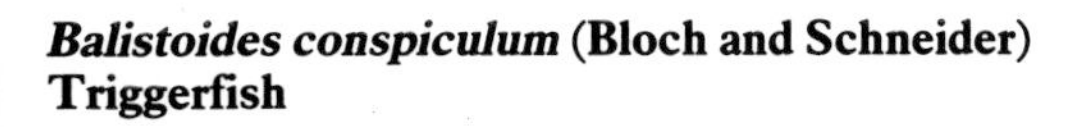

Balistoides conspiculum **(Bloch and Schneider)**
Triggerfish

Length 30 cm (12 inches). Found in the tropical Pacific, from Polynesia to Madagascar, China and Japan.

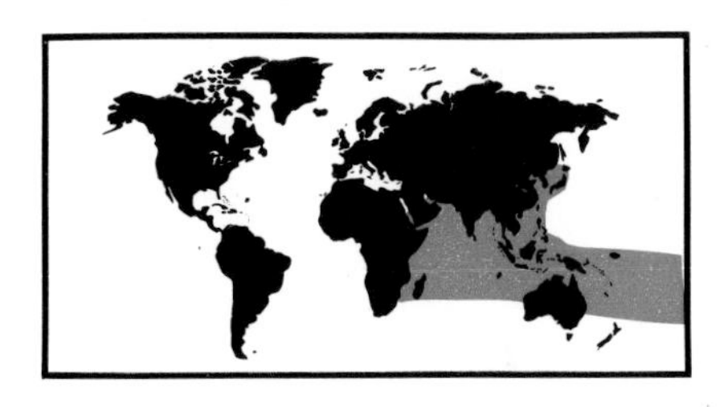

Family CARANGIDAE

Jacks, scads, and pompanos comprise a large group of swift-swimming oceanic fishes which are cosmopolitan in distribution. They are particularly common in the vicinity of coral reefs. Some of the Pacific species are noted especially for long migrations in quest of food. They are mostly carnivorous. Many are desirable game fishes, and most are valued as food. There are about 25 genera, with approximately 140 species.

Caranx hippos **(Linnaeus)**
Jack

Length 25 cm (10 inches). Lives in the tropical Atlantic.

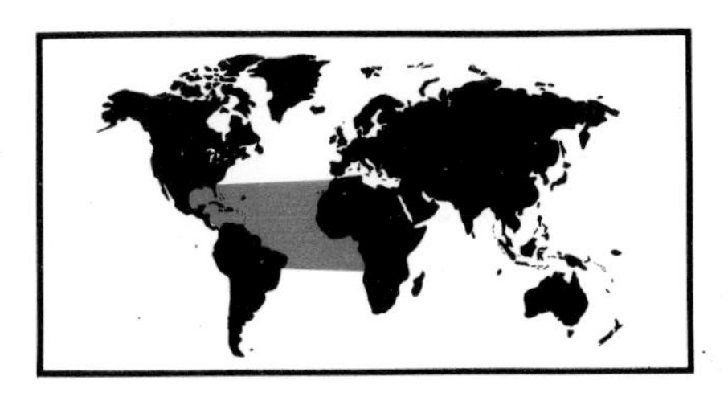

Caranx melampygus **Cuvier**
Green jack

Found in the tropical Pacific, north to Japan.

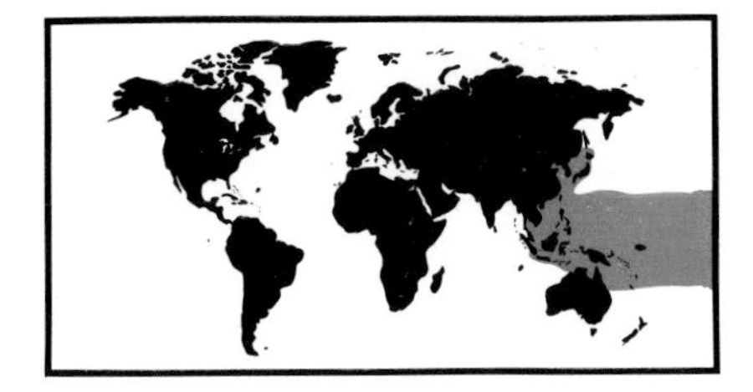

Family CLUPEIDAE

Herrings are one of the most widespread groups of fishes. Some species live in fresh water, while others enter rivers to spawn, but the majority are oceanic species. Herrings are frequently found swimming in immense schools. They have been implicated in both ciguatera and clupeoid types of ichthyosarcotoxism. Two species, *Clupanodon thrissa* and *Clupea tropica*, have been reported as violently toxic. Although these two species are considered to be valuable food fishes in most areas, they may be extremely poisonous at times. The dentition of herring is small and feeble, and their food consists largely of planktonic organisms, shrimp, crustaceans, worms and other small organisms. They are greatly valued as food fishes. The economic significance of herrings was noted by Jordan in 1905: "As salted, dried, or smoked fish the herring are found throughout the civilized world, and its spawning and feeding-grounds have determined the location of cities." There are perhaps 50 genera, with about 180 species in all.

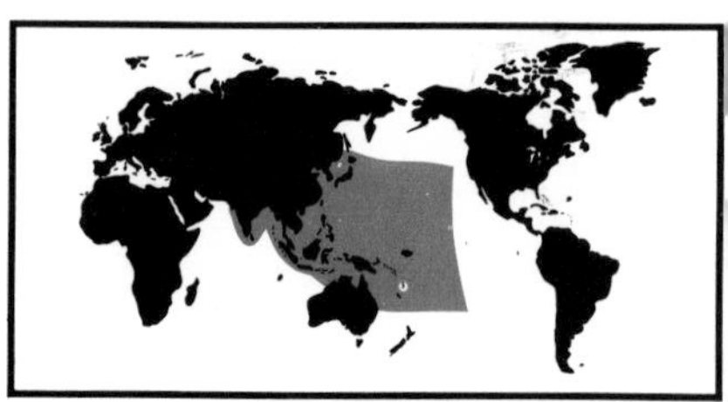

***Clupanodon thrissa* (Linnaeus)**
Herring

Length 25 cm (10 inches). Ranges throughout the waters of the tropical Pacific, Japan, China, Taiwan, Korea, Indonesia and India.

Family ENGRAULIDAE

Anchovies are small herring-like fishes, abundant in temperate and warm seas. Although generally found in the open sea, they also enter bays and ascend rivers. They are characterized by a snout which projects beyond the very wide mouth. The flesh of some species is of excellent flavor, and they are commercially canned in large numbers. Anchovies are valuable bait and serve as an important food source for large fishes. They are mainly plankton feeders. This family contains 16 genera, with about 139 species.

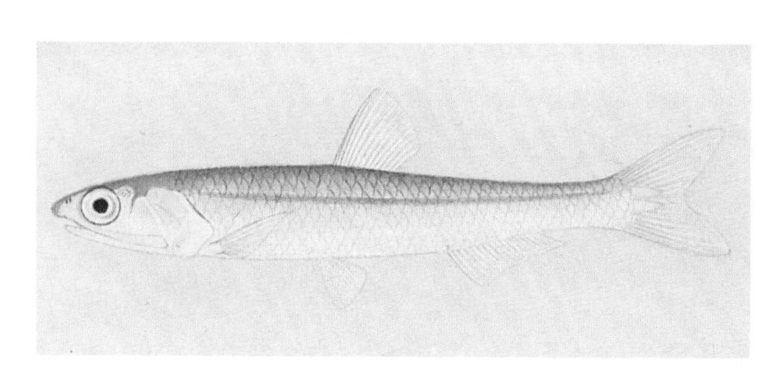

***Engraulis japonicus* (Schlegel)**
Anchovy

Length about 14 cm (5.5 inches). Found around China, Japan, Korea and Taiwan.

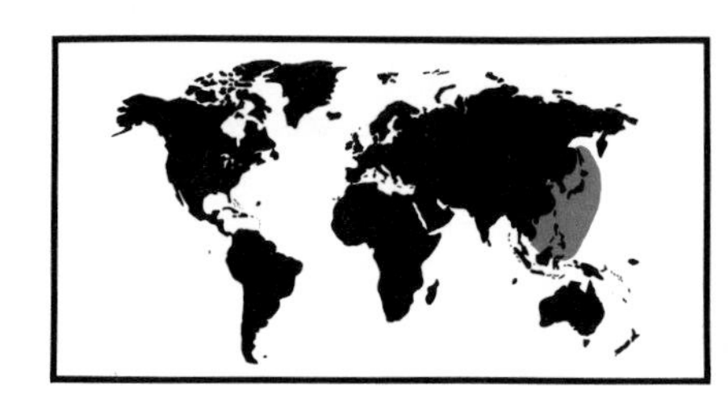

Family HOLOCENTRIDAE

The squirrelfishes, or soldierfishes, are characteristic species of rocky or coral regions in tropical seas. Their ctenoid scales are very hard and spiny. They are further characterized by the presence of a spine at the angle of the gill covering. The coloration of these fishes is usually a brilliant crimson, with or without stripes. Squirrelfishes are active predators and largely nocturnal in their habits, feeding on crustaceans, worms and algae. During the day they may be observed hovering almost motionless in coral or rock crevices. At night they feed on the reef flat or in the surf. This family consists of eight genera, with about 61 species.

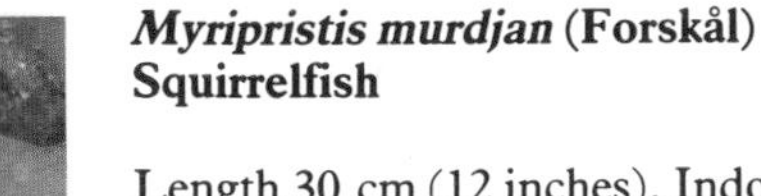

Myripristis murdjan **(Forskål)**
Squirrelfish

Length 30 cm (12 inches). Indo-Pacific.

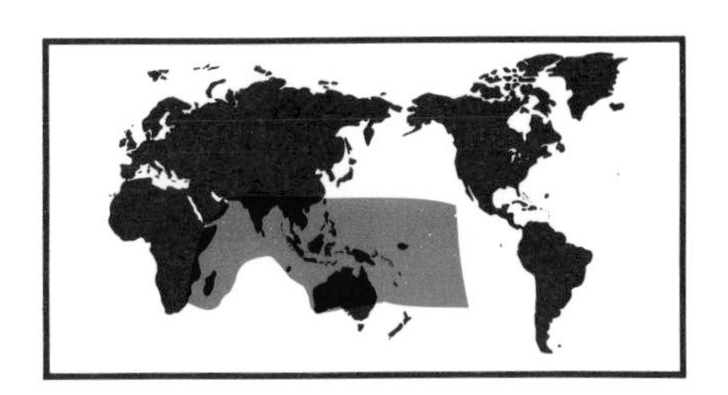

Family LABRIDAE

Wrasses are characterized by large, separate conical teeth in the front of the jaws. They have an extensive geographical range, but most are concentrated in warmer water. Most live in shallow coastal waters in rocky areas, coral reefs, and amongst growths of marine algae. Some species are herbivorous, but most are carnivores. They are among the most beautiful and gaily colored of the reef creatures. Their quality as food fishes varies considerably between species. There are about 57 genera, comprised of approximately 500 species.

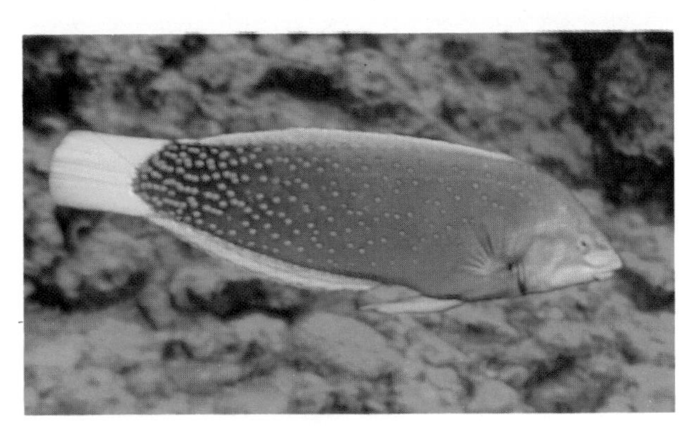

Coris gaimardi **(Quoy and Gaimard)**
Wrasse

Length 30 cm (12 inches). Tropical Indo-Pacific.

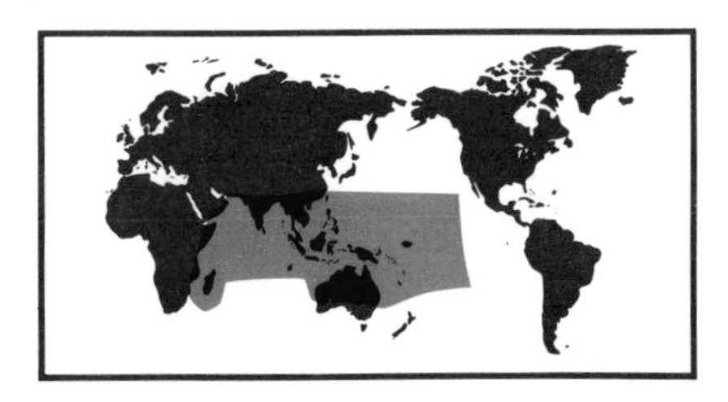

***Epibulus insidiator* (Pallas)**
Wrasse

Length 30 cm (12 inches). Tropical Indo-Pacific.

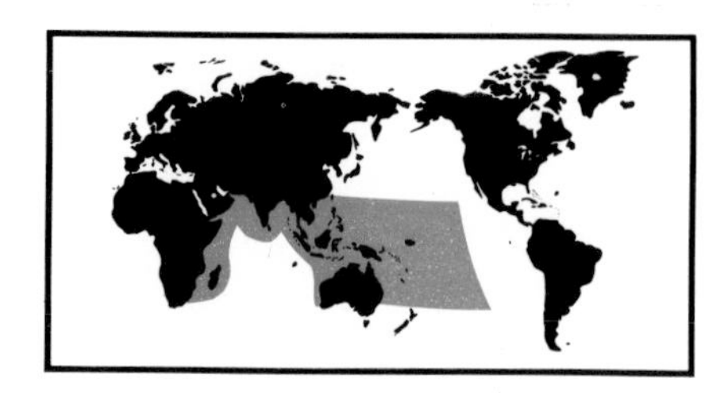

Family LUTJANIDAE

Snappers are carnivorous shore fishes, abundant in all warm seas. Their food consists mainly of smaller fishes. They are common in rocky, coral reef areas. Snappers are usually good food fishes, and of commercial value in some places. Certain members of the genus *Lutjanus* are known to accumulate ciguatera toxins. There are 17 genera, with an estimated 185 species.

***Aprion virescens* Valenciennes**
Snapper

Length 70 cm (27 inches). Tropical Indo-Pacific.

***Gnathodentex aureolineatus* (Lacépède)**
Snapper

Length 25 cm (10 inches). Tuamotu Archipelago, westward to east Africa.

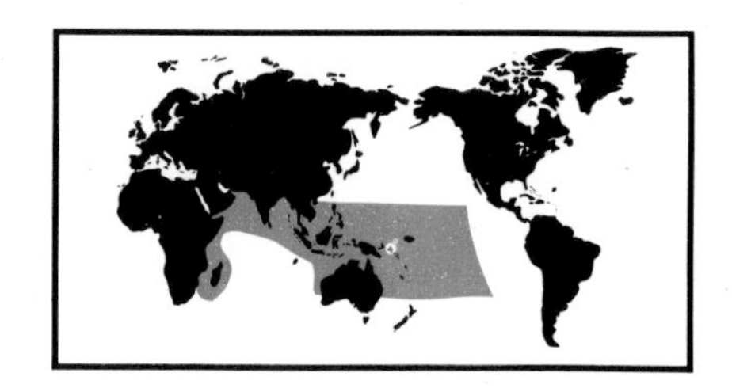

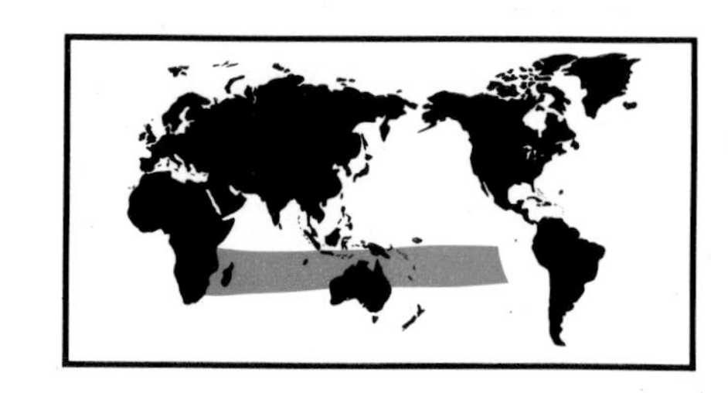

***Lethrinus miniatus* (Forster)**
Snapper

Length 45 cm (18 inches). Polynesia, westward to east Africa.

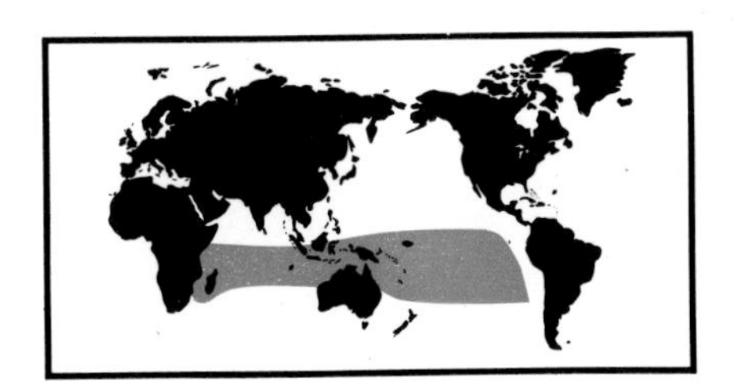

***Lutjanus bohar* (Forskål)**
Red snapper

Length 90 cm (35 inches). Tropical Pacific to east Africa and the Red Sea.

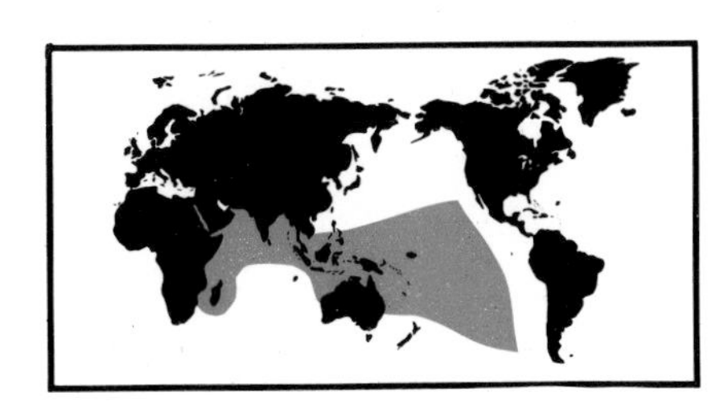

***Lutjanus monostigma* (Cuvier)**
Snapper

Length 30 cm (12 inches). Polynesia, westward to the Red Sea and China.

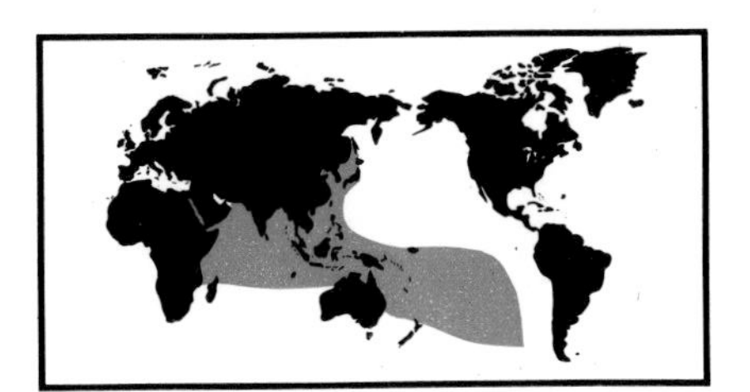

Lutjanus nematophorus **(Bleeker)**
Chinaman fish

Length 30 cm (12 inches). Australia.

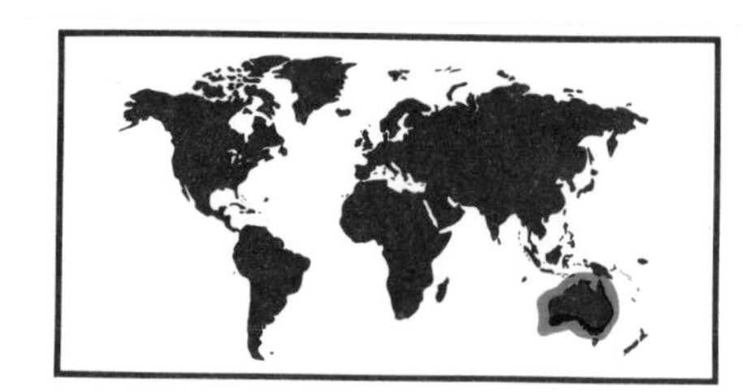

Lutjanus vaigiensis **(Quoy and Gaimard)**
Variegated snapper

Length 50 cm (20 inches). Polynesia, westward to east Africa and Japan.

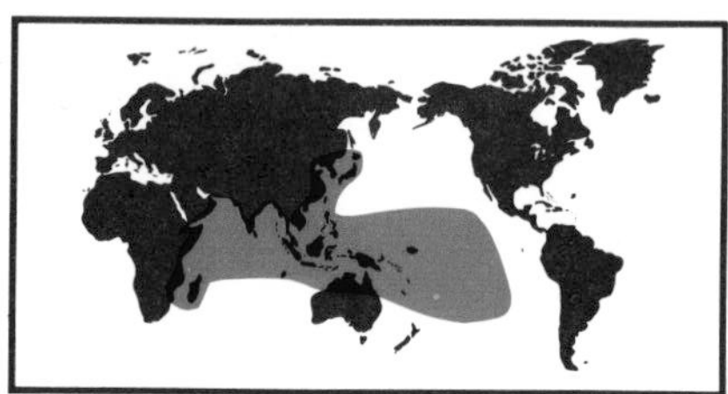

Monotaxis grandoculis **(Forskål)**
Snapper

Length 30 cm (12 inches). Polynesia, westward to east Africa.

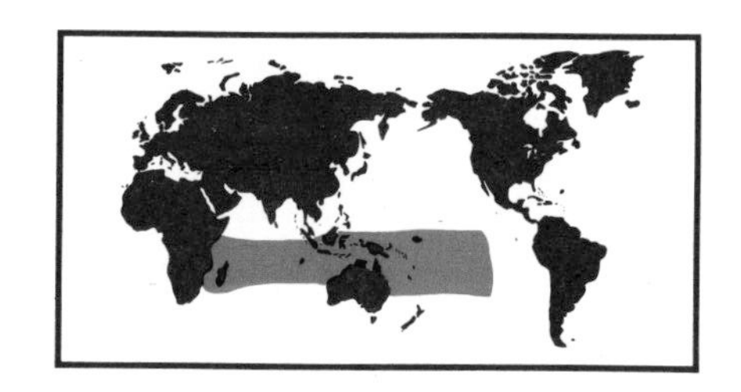

Family MULLIDAE

Surmullets, or goatfishes, are all small to moderate-sized shore fishes of warm seas. A variety of brightly-colored forms may be seen swimming about coral reefs. They are usually observed feeding in small schools, using barbels on their chins along the sandy areas. They are carnivorous, and feed largely on a variety of predatory fishes as their flesh is usually excellent. This family consists of approximately 55 species in six genera.

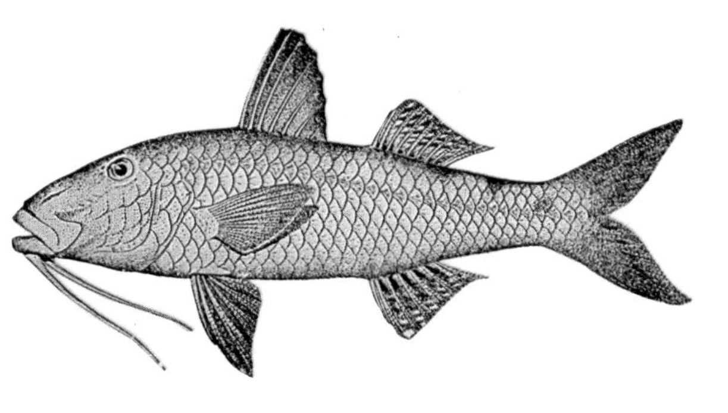

Parupeneus chryserydros **(Lacépède)**
Surmullet, goatfish

Length 30 cm (12 inches). Polynesia and Micronesia.

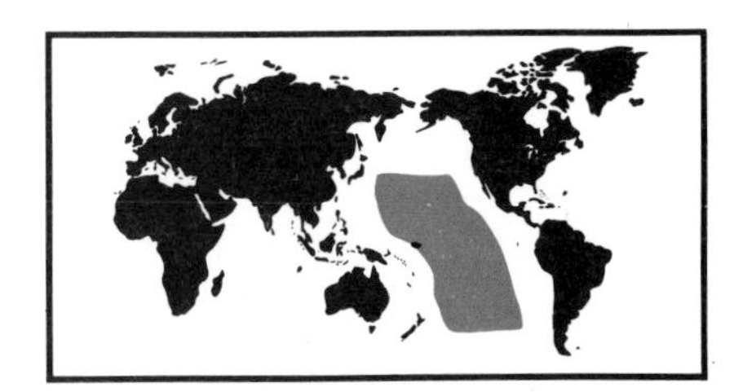

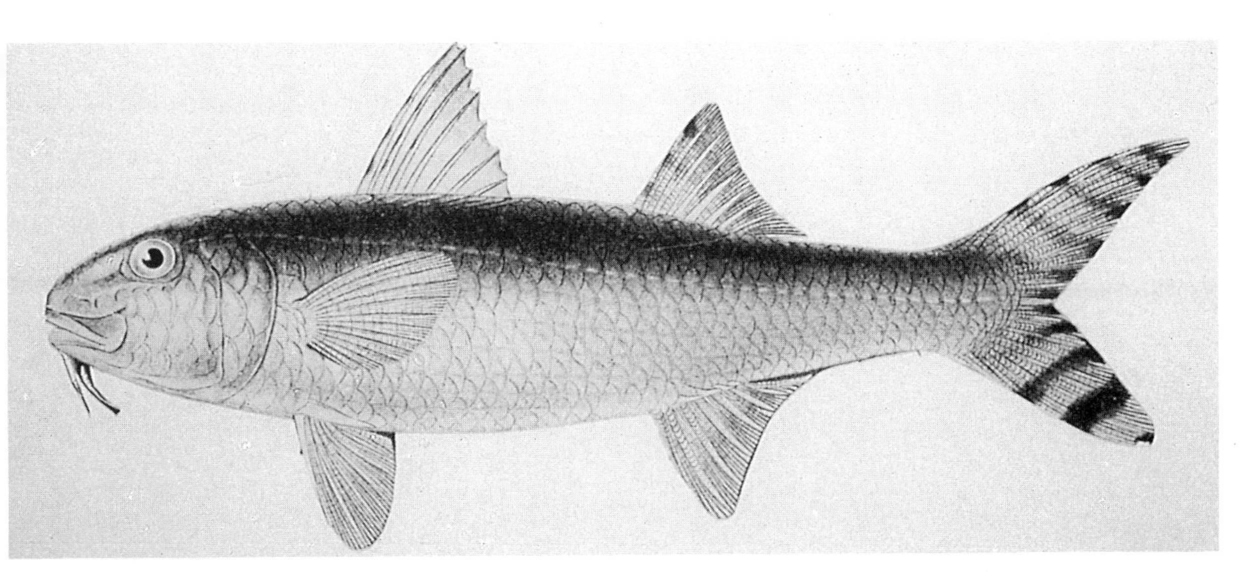

Upeneus arge **Jordan and Evermann**
Surmullet, goatfish

Length 30 cm (12 inches). Polynesia and Micronesia.

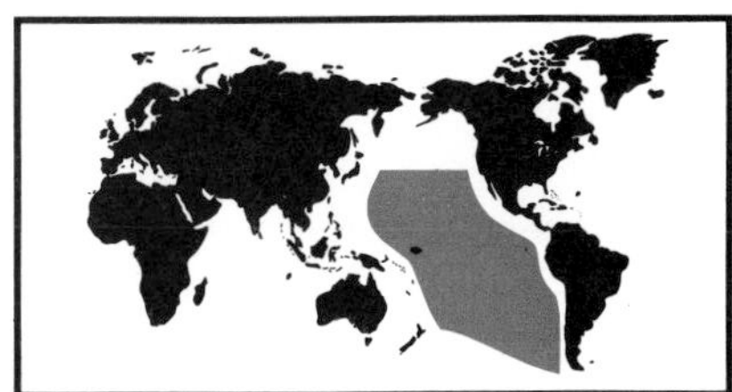

Family MURAENIDAE

Moray eels are abundant in tropical and semi-tropical seas. They are found from intertidal reef flats to depths of several hundred feet throughout temperate and tropical seas. Moray eels inhabit a large variety of ecological niches: surge channels, coralline ridges, inter-islet channels, reef flats, and lagoon patch reefs. They are nocturnal in habit and hide in crevices, holes and under rocks or coral during the day. With the aid of an artificial light, one can sometimes observe moray eels wriggling over a reef flat at night in large numbers. They are carnivorous in diet and may often be seen thrusting their heads out of coral holes, mouths slowly opening and closing, waiting for an unwary victim. They are able to strike with great speed. The long, fang-like, and depressible teeth are exceedingly sharp and can inflict serious lacerations. Their powerful muscular development, tough leathery skin, and dangerous jaws make them formidable adversaries. Some of the large morays attain a length in excess of 3 m (10 feet) and may constitute a real hazard to divers, particularly those who poke hands into holes and crevices. Morays can be readily lured with dead fish. Natives use spears, hooks and lines, snares, and traps to capture them. The flesh of some species is said to be agreeable, but is more often oily and not readily digestible. Some species are poisonous. There are about 100 species in 12 genera.

Gymnothorax flavimarginatus **(Rüppell)**
Moray eel

Hawaiian Islands, westward to east Africa.

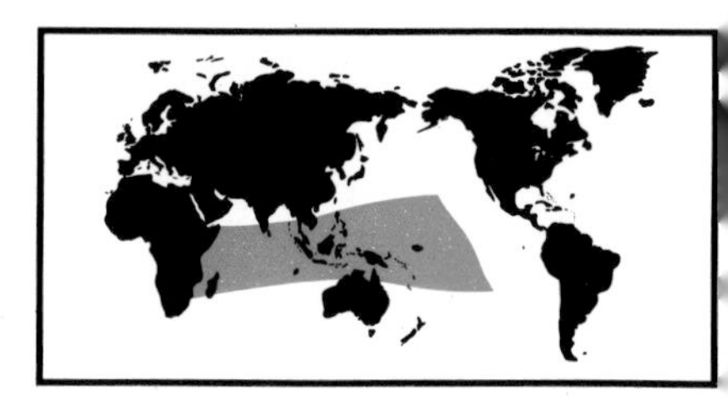

Gymnothorax javanicus **(Bleeker)**
Moray eel

Hawaiian Islands, westward to east Africa.

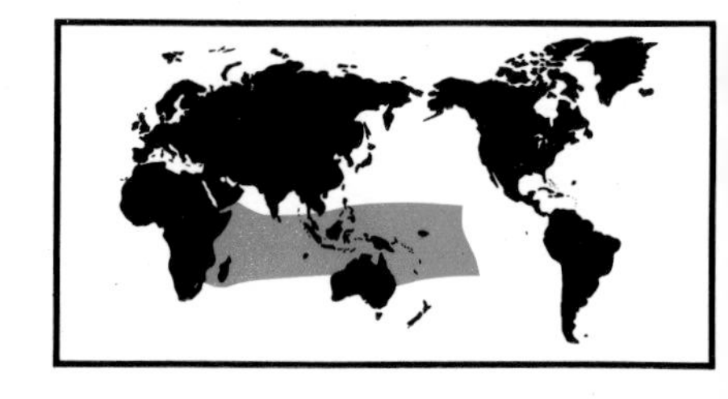

Gymnothorax meleagris **(Shaw and Nodder)**
Moray eel

Hawaiian Islands, westward to east Africa.

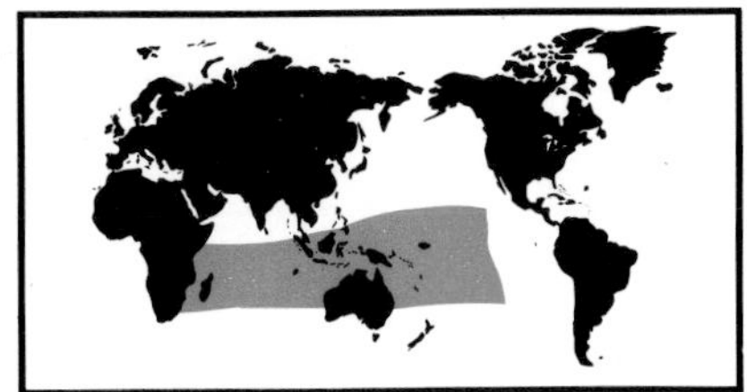

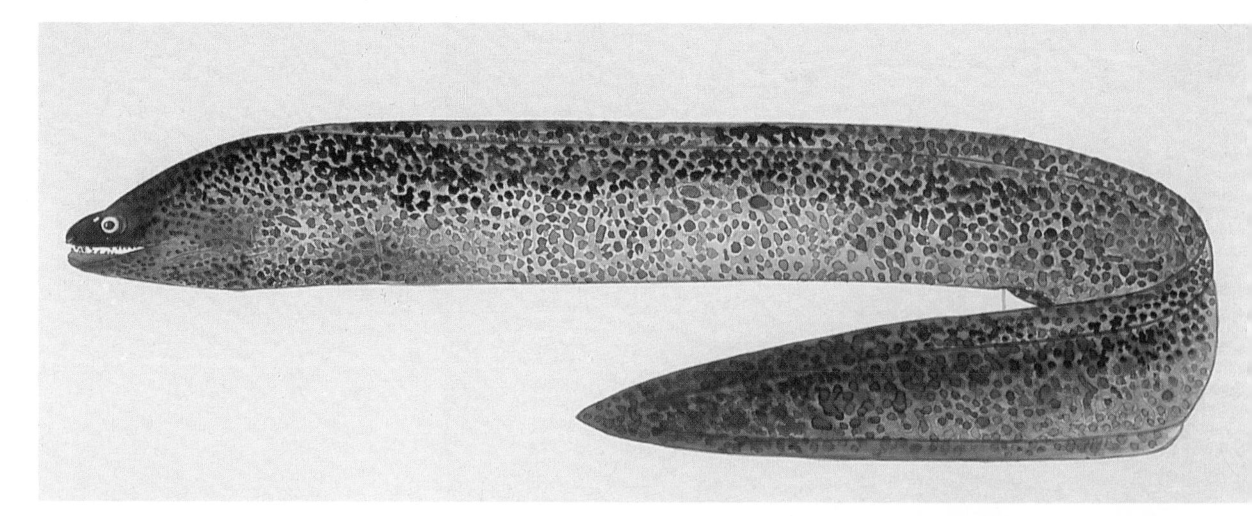

Gymnothorax pictus **(Ahl)**
Moray eel

Polynesia to east Africa.

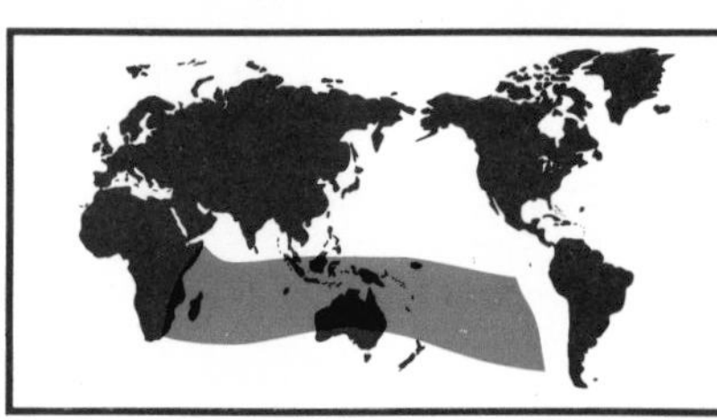

Gymnothorax undulatus **(Lacépède)**
Moray eel

Hawaiian Islands to the Red Sea and east Africa.

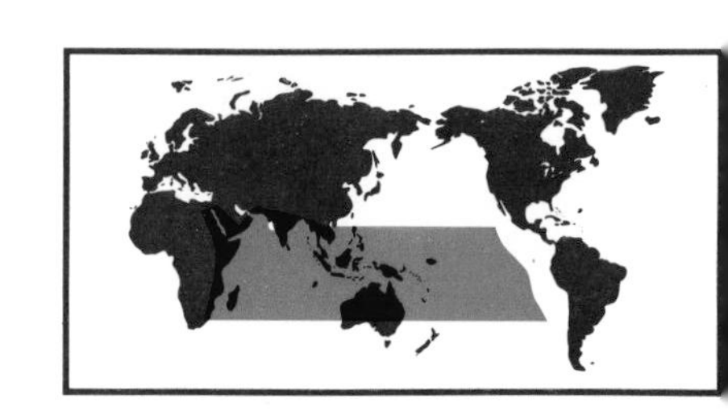

Family OSTRACIIDAE

Trunkfishes, or cowfishes or boxfishes, comprise one of the more peculiarly constructed groups of fishes, named because the body is enveloped within a bony box consisting of six-sided scutes. They live in tropical seas, and are frequently seen swimming slowly along among the corals in shallow water. Many are brilliantly colored. The "exoskeleton", which serves them well for protection, is also used by primitive peoples as a container for cooking them over a fire. Little is known of their habits. In all there are some 30 species which are included in about 13 genera.

Lactophrys trigonus **(Linnaeus)**
Trunkfish

Length 30 cm (12 inches). Atlantic coast of tropical America, north to Cape Cod.

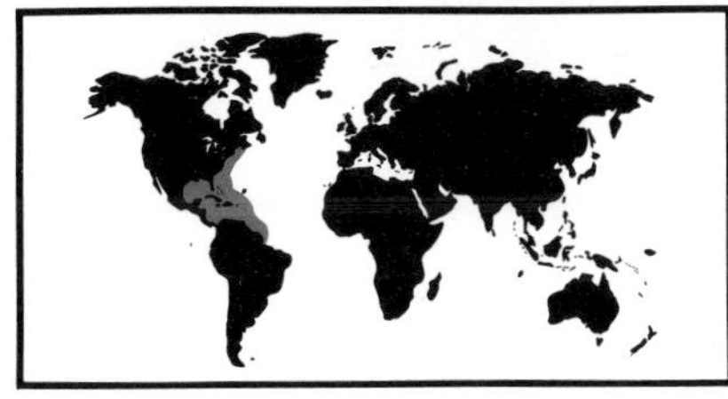

Lactoria cornuta **(Linnaeus)**
Trunkfish

Tropical Pacific.

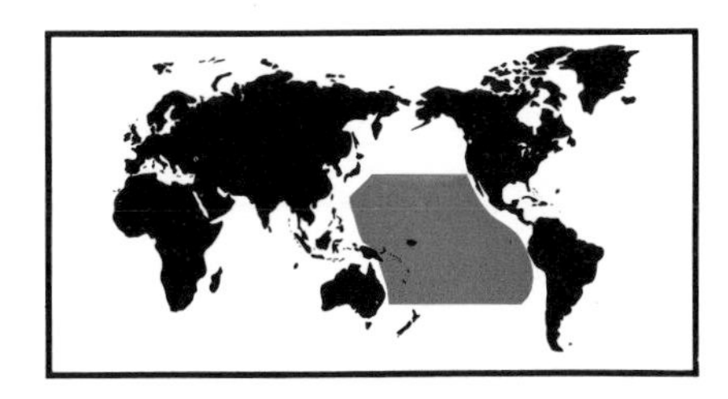

Family SCARIDAE

Parrotfishes are similar in appearance to wrasses, but differ in that they have teeth fused into plates. They are shallow water shore fishes and very common in coral reefs, lagoons and rocky areas of tropical waters. Their food consists of algae and small animals that are ingested as they graze upon coral rubble. Deep gouges and scratches may be observed on coral rocks where parrotfishes have browsed. In this way they contribute significantly to the formation of fine sand, returning the pulverized rock and skeletal material to the bottom in their faeces. There are about 11 genera, with perhaps 64 species.

***Scarus coeruleus* (Bloch)**
Blue parrotfish

Length 90 cm (36 inches). Florida and the West Indies.

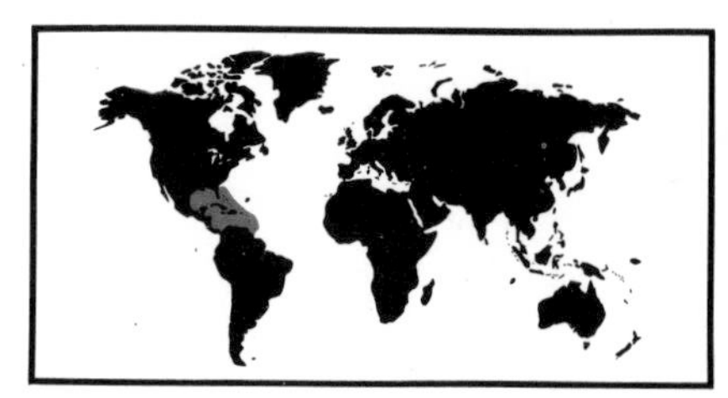

***Scarus microrhinos* (Bleeker)**
Parrotfish

Length 30 cm (12 inches). Indo-Pacific.

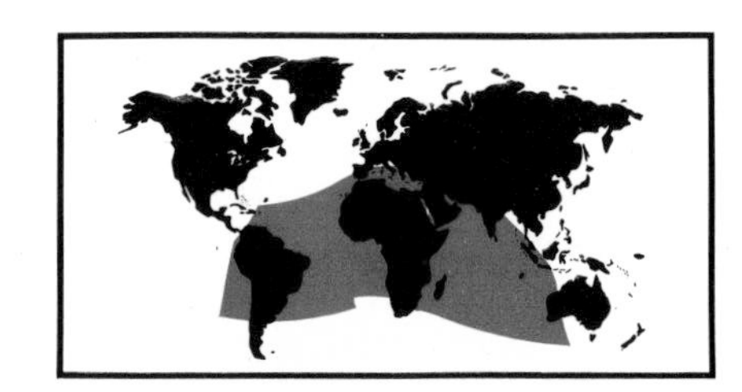

Family SCOMBRIDAE

Members of the family of true mackerels, which includes the tunas, are distinguished by their streamlined bodies, smooth scales, metallic coloration, and the presence of a number of detached finlets behind the dorsal and anal fins. Mackerels are for the most part swift-swimming and schooling pelagic fishes and are one of our most valuable fisheries resources. Some schools have been reported to be "a half mile wide and 20 miles long" (equivalent to 800 m wide and 32 km long). Although there are a large number of species in the Scombridae, only a few members of the genera *Acanthocybium, Euthynnus, Sarda,* and *Scomberomorus* have been incriminated in ciguatera poisoning. They, like most of their relatives, are oceanic fishes, but during the reproductive season come in close to shore; some species may at times be found in bays and lagoons. Apparently it is during this inshore migration that they acquire ciguatoxin. This family of true mackerels consists of 48 species included within 15 genera.

***Euthynnus pelamis* (Linnaeus)**
Oceanic bonito

Length 50 cm (20 inches). Found in tropical areas around the world.

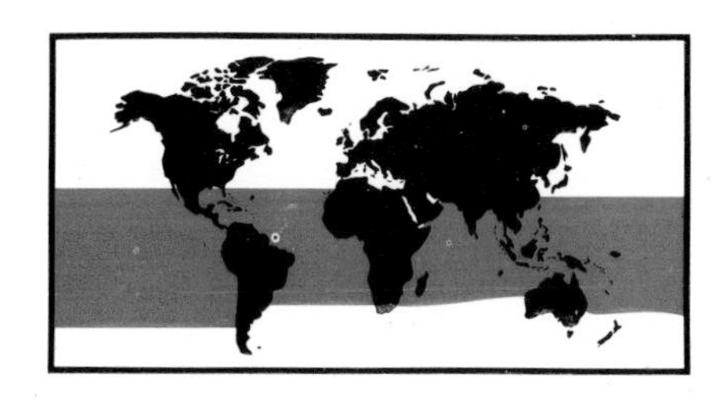

Family SERRANIDAE

Seabass or groupers are robust, carnivorous, and predatory shore fishes of tropical and temperate waters. A variety of biotopes are inhabited by this large family: coral reefs, rocks, sandy areas and kelp. A few are found in fairly deep water, but most live in shallow water. Some attain great size. They are usually considered to be good food fishes. This family comprises approximately 35 genera with about 370 species.

***Cephalopholis argus* (Bloch and Schneider)**
Seabass, grouper

Length 50 cm (20 inches). Tropical Indo-Pacific.

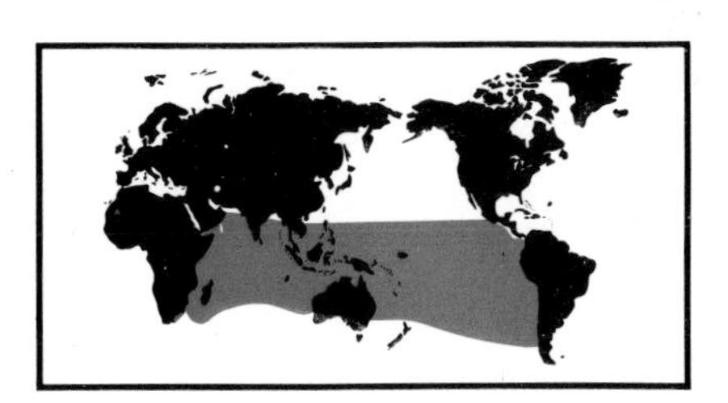

***Epinephelus fuscoguttatus* (Forskål)**
Seabass, grouper

Indo-Pacific.

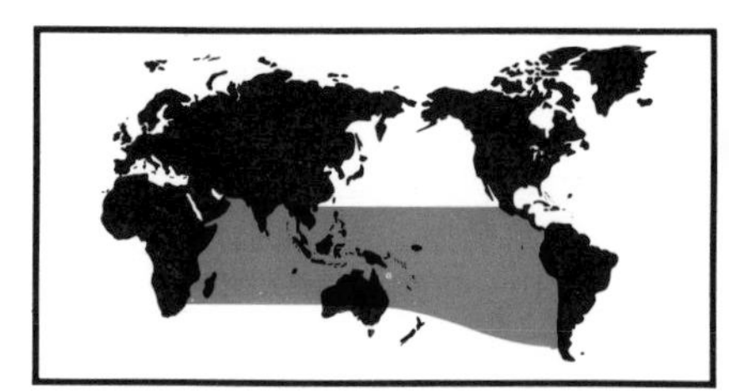

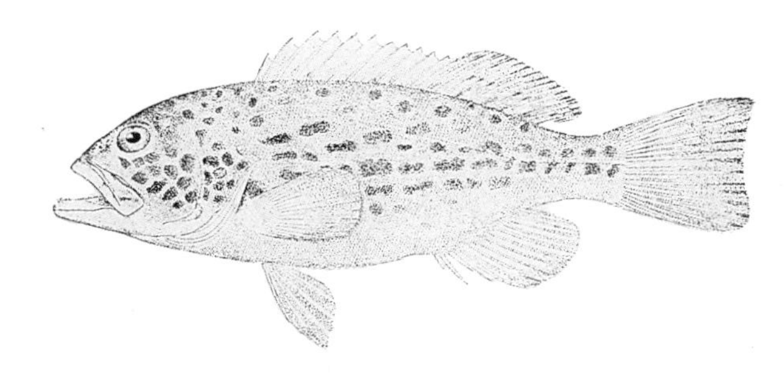

***Mycteroperca venenosa* (Linnaeus)**
Seabass

Length 90 cm (35 inches). Western tropical Atlantic.

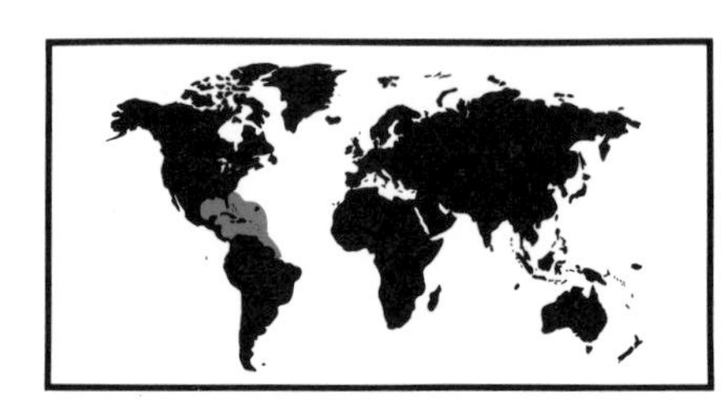

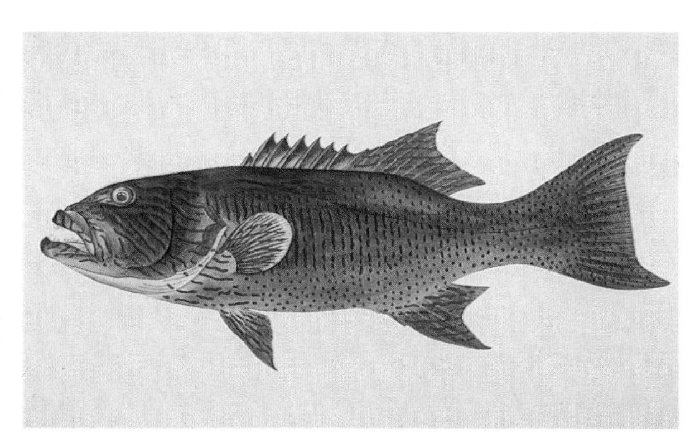

***Plectropomus obligacanthus* (Bleeker)**
Seabass

Length 56 cm (22 inches). Indonesia, the Philippines, the Caroline and Marshall Islands.

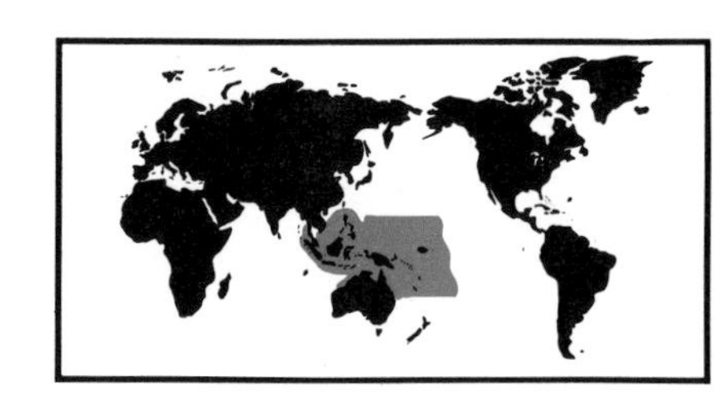

Plectropomus truncatus **(Fowler)**
Seabass

Length 50 cm (20 inches). Micronesia, Indonesia and the Philippines.

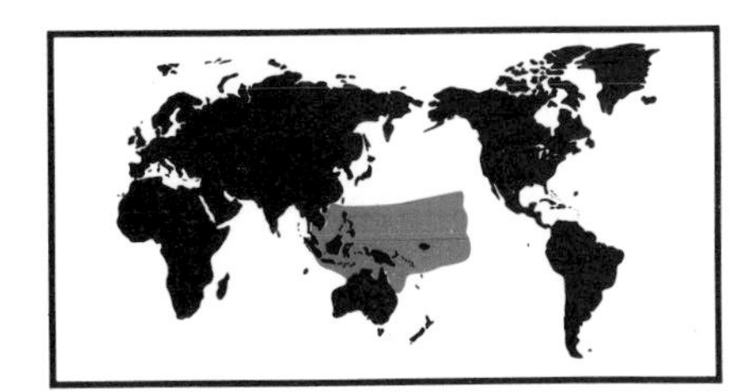

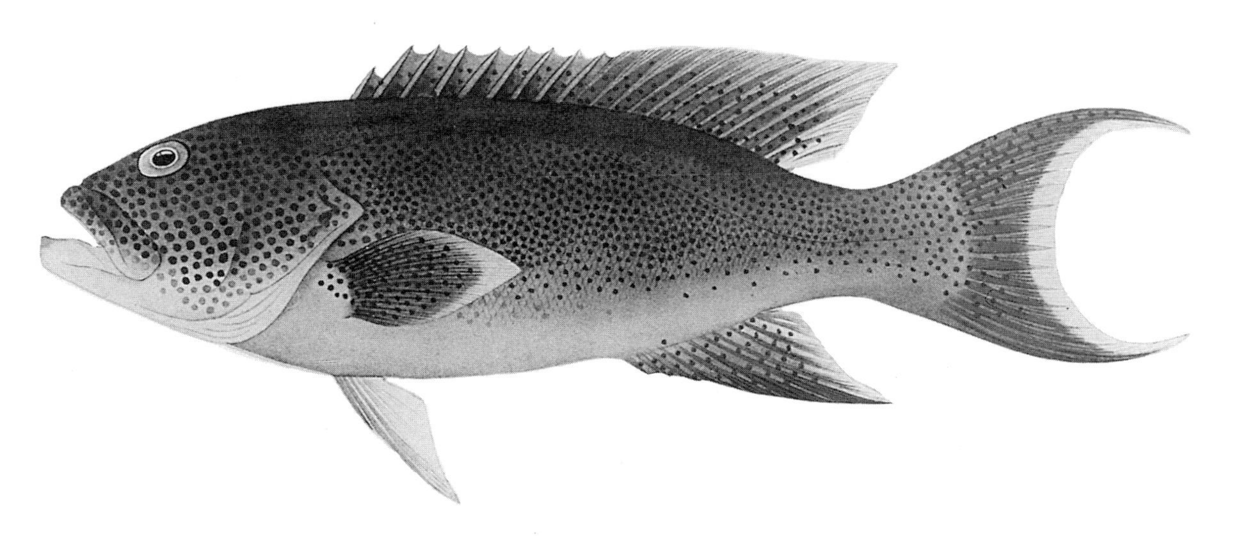

Variola louti **(Forskål)**
Seabass

Length 60 cm (24 inches). Tropical Indo-Pacific.

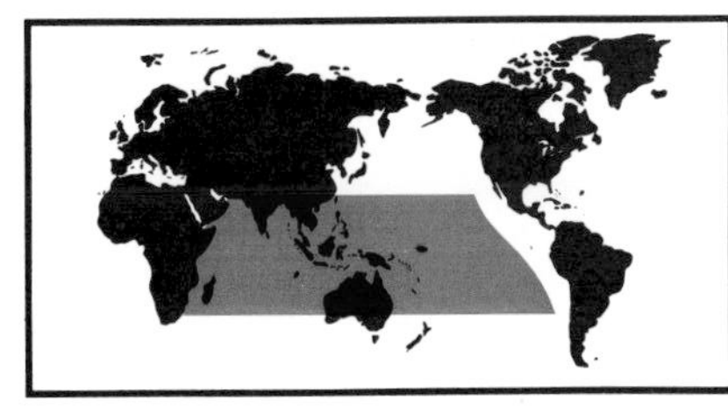

Family SPARIDAE

Porgies are a group of shore fishes having a perch-like appearance. They are found along tropical and temperate coasts, usually confined to shallow waters in a variety of habitats. Small fishes, crustaceans, and other invertebrates make up their diet. They are among the more important food fishes. There are about 100 species included in 29 genera within this family.

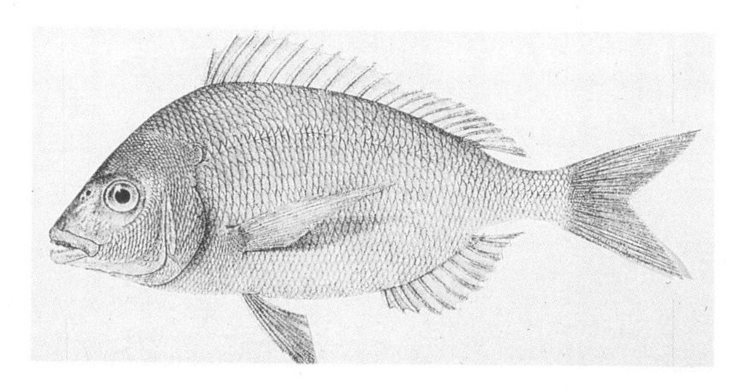

***Pagellus erythrinus* (Linnaeus)**
Porgie

Length 40 cm (16 inches). Black Sea, Mediterranean, and east Atlantic, from the British Isles and Scandinavia to the Azores, the Canaries and Fernando Poo (Gulf of Guinea). The toxic nature of this species is unclear.

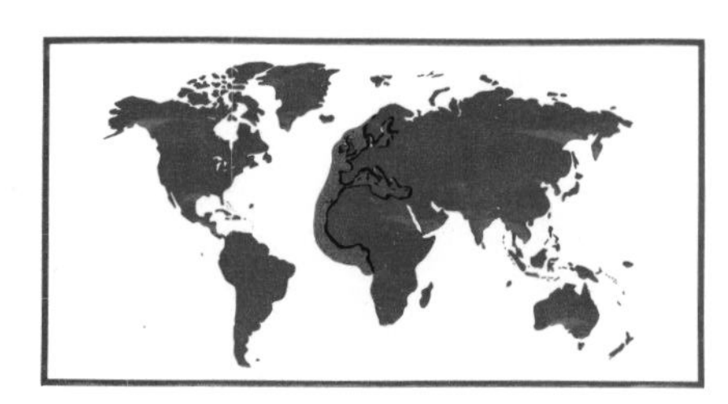

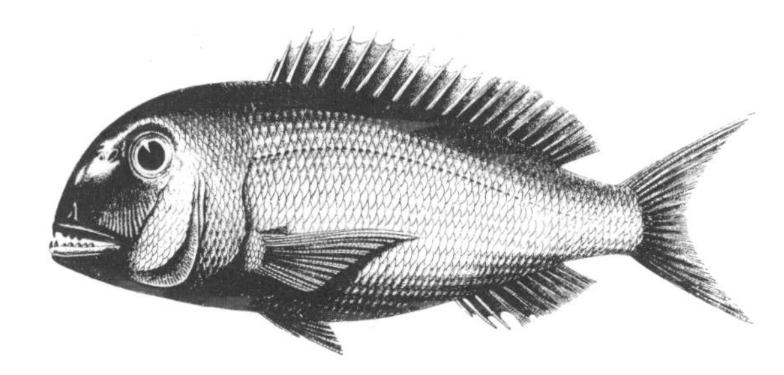

***Pagrus pagrus* (Linnaeus)**
Porgie

Length 40 cm (16 inches). Eastern Atlantic and the Mediterranean Sea. Its toxic nature is uncertain.

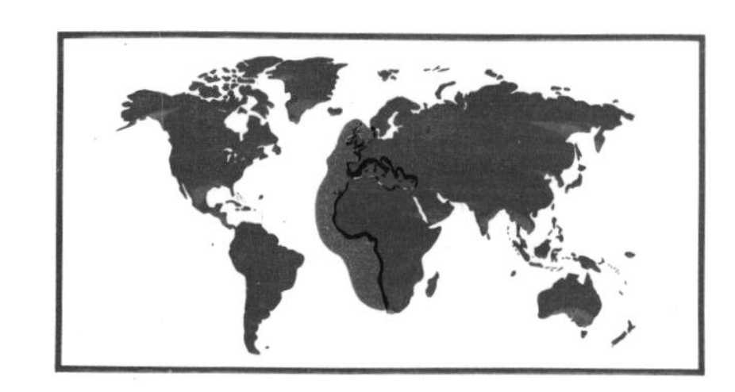

Family SPHYRAENIDAE

Barracuda are long, slender, and swift-swimming carnivores of tropical and temperate waters. Their jaws contain long, knife-like and canine teeth. They are common in lagoons, passageways and coral reefs. Their flesh is of excellent quality, but the large specimens of some species may become toxic. There is one genus, with 18 species in this family.

***Sphyraena barracuda* (Walbaum)**
Barracuda

Length up to 1.5 m (5 feet). Indo-Pacific, from Hawaii to the Red Sea, and the west Atlantic from Brazil to the West Indies, Florida and Bermuda. Specimens over 9 kg (20 pounds) in weight are suspect.

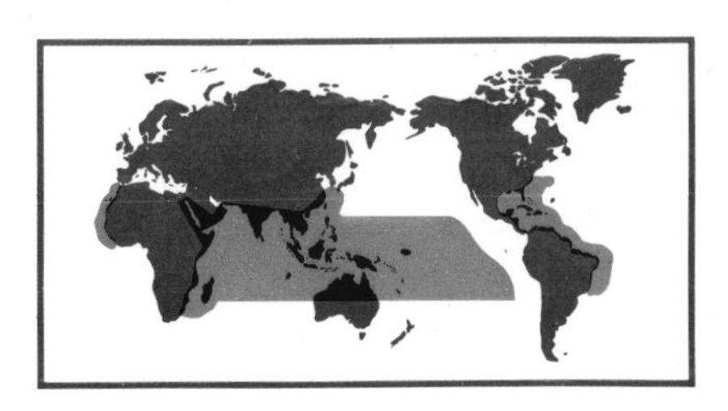

Family TETRAGONURIDAE

The squaretails have bodies which are slender and rounded. They are primarily oceanic fishes found in tropical, subtropical, and temperate seas. The younger squaretails live among the medusae and other planktonic animals. This family has three species within one genus.

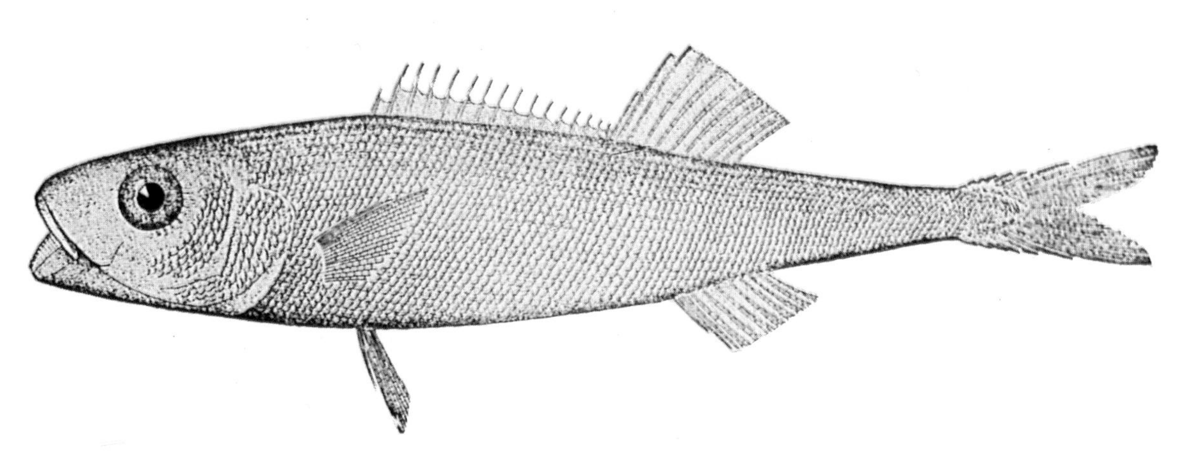

***Tetragonurus cuvieri* (Risso)**
Squaretail

Length 25 cm (10 inches). Throughout all temperate regions.

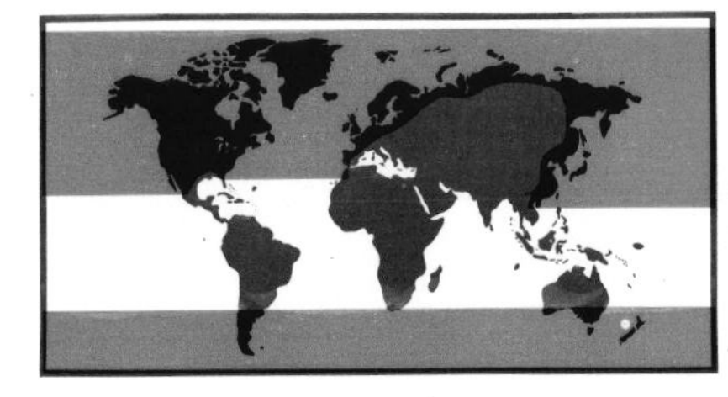

SCOMBROID FISH POISONING

This form of poisoning involves predominantly the scombroid fishes, which includes tuna, mackerel, skipjack, and related species. The fishes included within this category are usually safe to eat and are a valuable commercial food product. Scombroid poisoning is caused by improper preservation of the fish. The poisons, histamine and its variants, develop in the flesh of the fish through bacterial activity on the musculature. For unknown reasons the histamine ingested within spoiled fish is much more toxic than chemical preparations of histamine ingested in an aqueous solution. This is believed to be due to the presence of potentiators of toxicity in spoiled fish. These potentiators may include the diamines cadavarine and putrescine. This is the only documented form of ichthyosarcotoxism in which bacteria (which decompose the fish) play an active role in the production of the poison within the body of the fish. However, the poison is not a traditional bacterial toxin. Since any of the tuna, skipjack, bonito, mackerel, etc may be involved, none are listed separately here.

GEMPYLID FISH POISONING

Gempylid poisoning results from eating the flesh of fishes of the family Gempylidae. The snake mackerels, as they are called, have elongated bodies that are laterally compressed and may reach a length of 2 m (6.5 feet) and weigh 75 kg (165 pounds). They are mostly found in schools and are caught in deep water at night. The species most implicated in poisoning is the castor oil fish (*Ruvettus pretiosus*), which ranges throughout the tropical Atlantic and Indo-Pacific Oceans. This species is normally taken by hook and line at night in depths of 730 m (400 fathoms) or more. When eaten, it is said to cause diarrhea. It may attain a length of 1.5 m (5 feet).

HALLUCINOGENIC FISH POISONING

Ichthyoallyeinotoxism, or hallucinogenic fish poisoning, is caused by ingesting certain types of reef fishes known to occur in the tropical Pacific and Indian Oceans. Several different types of tropical reef fishes can produce hallucinations after they are eaten. This kind of poisoning is difficult to predict.

Family GEMPYLIDAE

***Ruvettus pretiosus* (Cocco)**
Castor oil fish

Length up to 1.5 m (5 feet).

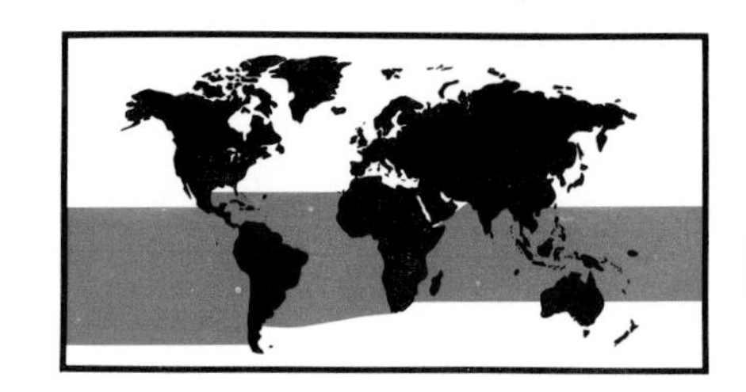

Family KYPHOSIDAE

The rudderfishes are characterized by the absence of molars, the front of the jaws being occupied by incisors which are often serrated, loosely attached and movable. There are numerous species, and they are largely inhabitants of warmer seas, although a few species are found in temperate waters. Rudderfish are usually found in shallow water around rocks, reefs or tidepool areas. They tend to be herbivorous in their eating habits. Seventeen genera with about 45 species make up this family.

Kyphosus cinerascens **(Forskål)**
Sea chub

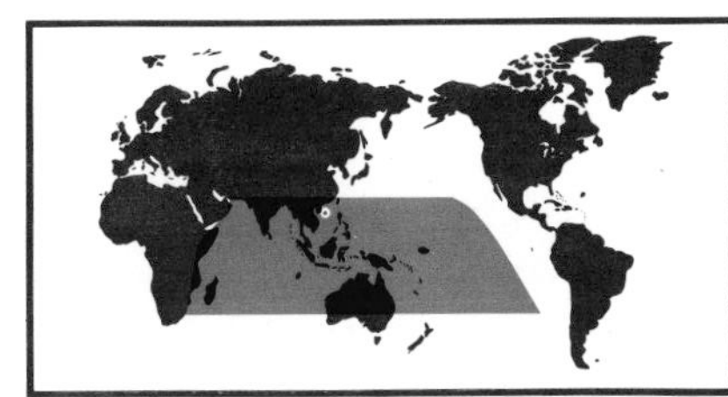

Length 50 cm (20 inches). Located in the Indo-Pacific region. These fish appear to be more commonly involved in hallucinogenic fish poisoning than most other related species.

Family MULLIDAE

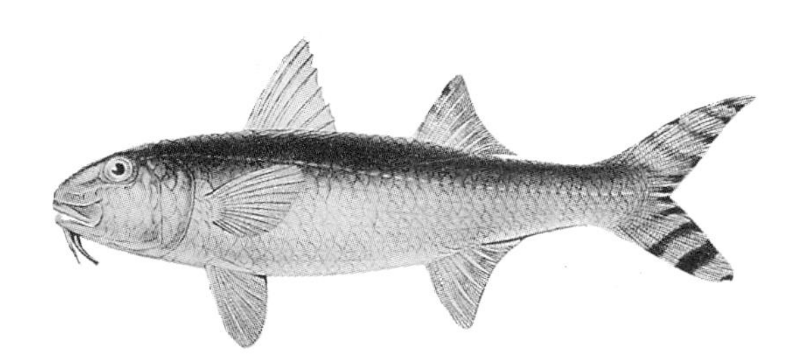

Upeneus arge **(Jordan and Evermann)**
Goatfish

Seems to be more often involved in this type of poisoning than other related species.

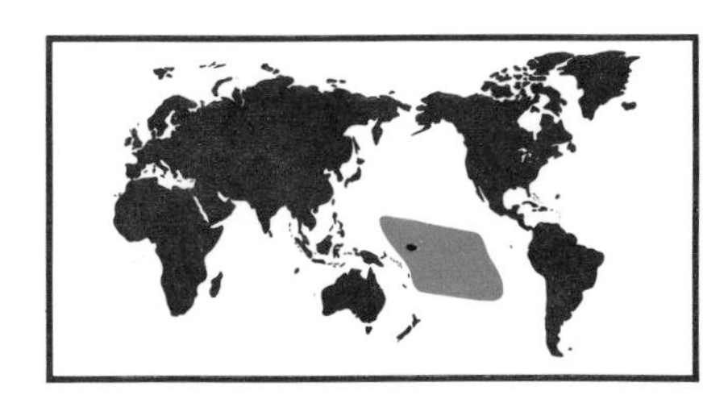

PUFFER POISONING

This type of fish poisoning is one of the most serious forms of intoxication. Of all marine creatures, puffers are among the most poisonous. The liver, intestines and skin usually contain a potent poison, which may cause rapid death. Puffer poisoning can be caused when a person eats a puffer, porcupine fish, or ocean sunfish. Puffers are known by many names throughout the world, including globefish, blowfish, balloon fish, toadfish, swellfish, fugu, botete, fahaka, tinga, and several more. The puffer is known for its remarkable ability to blow itself up by gulping in a large amount of air or water. They are quite noisy during inflation because they grind their heavy jaws and teeth together. Some puffers inflict nasty bites.

Although highly poisonous, puffers are considered a delicacy in Japan. Called "fugu", they are prepared and sold in special restaurants where highly trained and experienced chefs prepare this fish for diners. The fugu is carefully prepared to make it safe to eat. Nevertheless, it is still the number one cause of fatal food poisoning in Japan, especially among people who fail to take the necessary precautions.

Tetrodotoxin elaborated by puffers causes a syndrome of neuromuscular paralysis in its most severe form. Death is usually due to respiratory failure. Milder intoxications cause light headedness and numbness and tingling, soon followed by gastrointestinal distress, weakness and low blood pressure. Because there is no specific antidote, treatment is supportive and based upon symptoms, similar to that for paralytic shellfish poisoning (see Chapter 7).

Family DIODONTIDAE

The porcupine fishes are spiny marine fishes which are widely distributed in all warm seas. Most diodontids range from 20 to 50 cm (8 to 20 inches) in length, inhabiting coral reefs and shoal areas. Porcupine fishes generally travel singly or in pairs. They have the classical ability to inflate themselves until they are almost spherical in outline. During the inflation process the spines which are usually depressed are extended, giving the fish a formidable appearance. Porcupine fishes have been known to kill larger carnivorous fishes by inflating themselves and becoming stuck in the throat of the would-be captor. Darwin (1945) stated that diodons had been found alive in the stomachs of sharks, and in some instances had mortally wounded their captors by gnawing through the stomach walls and sides of the sharks. This family includes two genera with a total of 15 species.

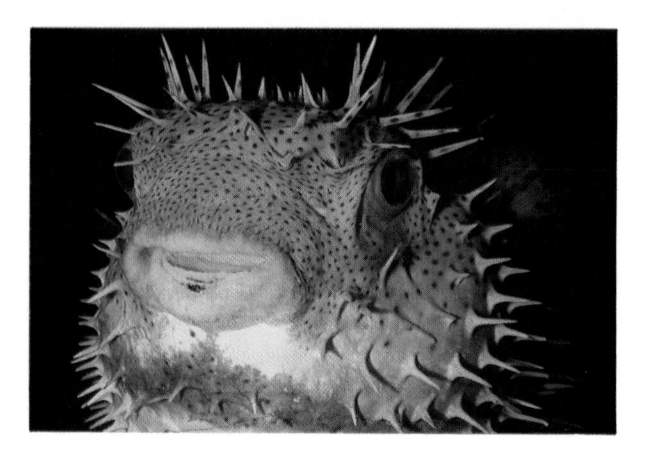

***Diodon histrix* Linnaeus**
Porcupine fish

Length up to 90 cm (35 inches). Found throughout all tropical waters, occasionally entering temperate areas.

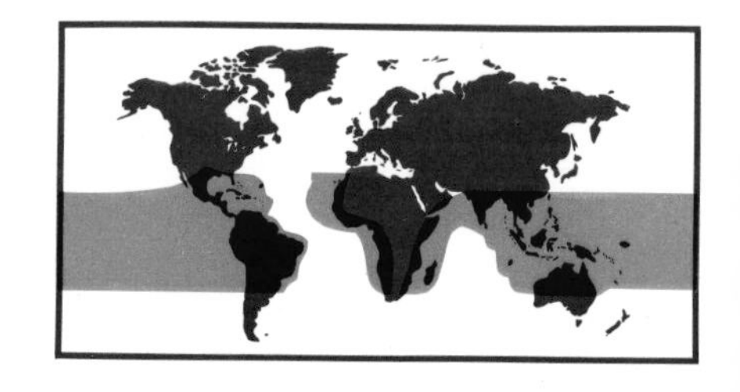

Family TETRAODONTIDAE

Tetraodontids inhabit a variety of ecosystems, including marine, estuarine, and fresh water. Most range in size from about 25 to 50 cm (10 to 20 inches), but *Lagocephalus lagocephalus* is said to attain a total length of 61 cm (24 inches), while some of the other puffers attain even greater sizes. *Arothron stellatus* has been reported to reach a total length of 91 cm (36 inches). Most puffers are considered to be shallow water fishes, but some members inhabit relatively deep waters. *L. oceanicus* has been taken at 40 m (130 feet) or more, *L. lagocephalus inermis* at 90 m (300 feet), *L. scleratus* at 60 m (200 feet), and *Sphaeroides oblongus* at 100 m (330 feet). *Liosaccus cutaneus* is usually considered to be a deep water species and *Boesmanichthys firmamentum* has been taken at a depth of 180 m (600 feet). Species of *Lagocephalus* have been taken hundreds of miles from shore at depths of 7300 m (4000 fathoms), and are a common constituent in the stomachs of pelagic fishes such as tunas, wahoos and other scombroids.

Whether these deep water puffers are nontoxic or the poison is not transvectored by scombroids is not known. The Indo-Pacific tetradontids living around coral reef areas tend to travel singly or in groups. *Sphaeroides* species are more gregarious and are sometimes observed in large groups. Stomach analyses conducted in laboratories on *A. hispidus* indicate that this species is omnivorous in its eating habits, taking fragments of corals, sponges, algae, mollusks and fish.

At Tagus Cove, Isabela Island, during the Kreiger Galapagos expedition (Halstead and Schall, 1955), it was observed that hundreds of *Sphaeroides annulatus* were attracted to the surface of the water with a night light. Specimens could then be readily captured by spear or dip net.

When puffers are at rest they appear to hover almost motionless with only their pectorals fanning the water. The pectoral, dorsal and anal fins are the chief locomotory organs, the tail being used principally as a rudder. There are 16 genera with about 118 species comprising this family of puffer fish.

Arothron hispidus **(Linnaeus)**
Maki-maki or deadly death puffer

Length 50 cm (20 inches). Panama Canal throughout the tropical Pacific to South America, Japan and the Red Sea.

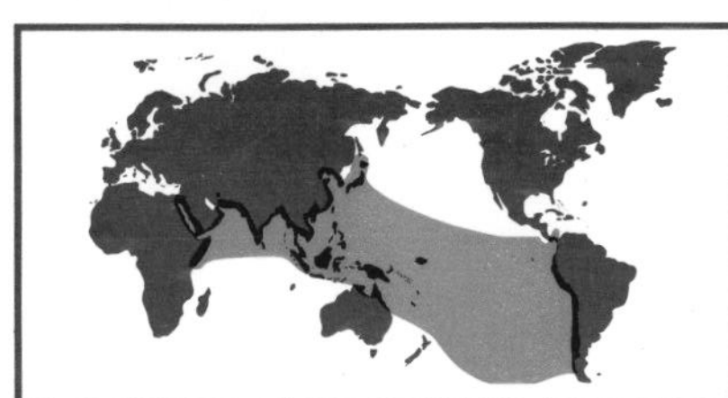

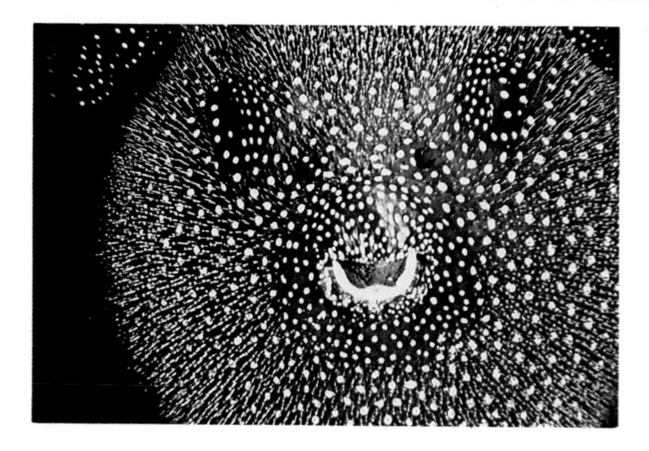

Arothron melagris **(Lacépède)**
White-spotted puffer

Length 33 cm (13 inches). West coast of Central America to Indonesia.

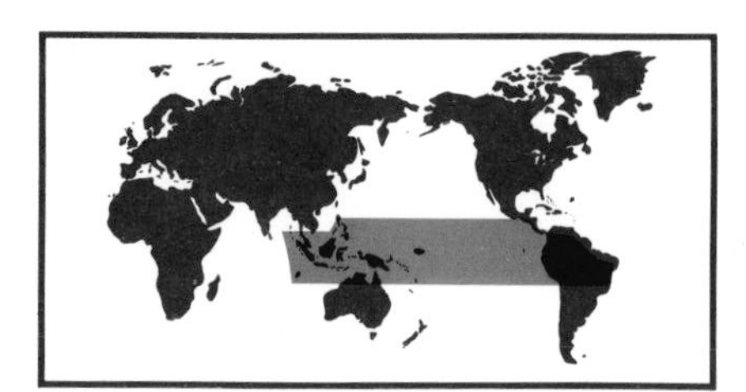

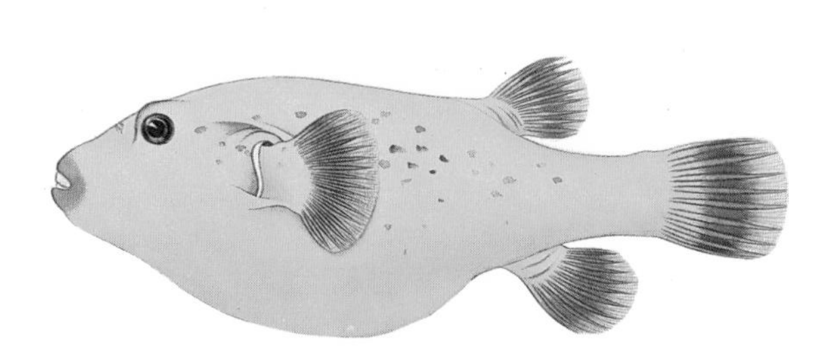

Arothron nigropunctatus **(Bloch and Schneider)**
Black-spotted puffer

Length up to 25 cm (10 inches). Polynesia, tropical Indo-Pacific to east Africa and the Red Sea.

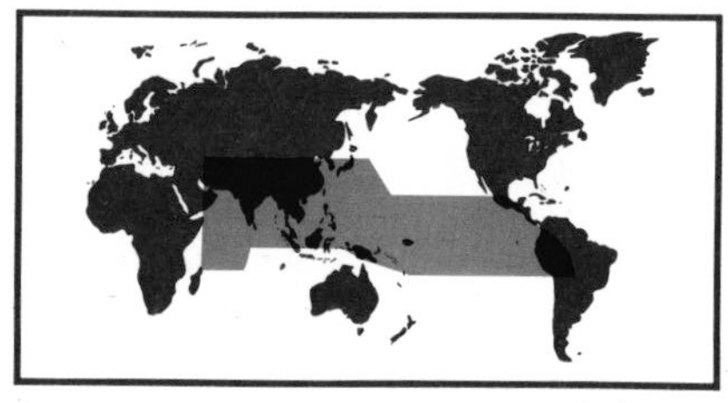

Sphaeroides annulatus **(Jenyns)**
Gulf puffer

Length 25 cm (10 inches). California to Peru, and the Galapagos Islands.

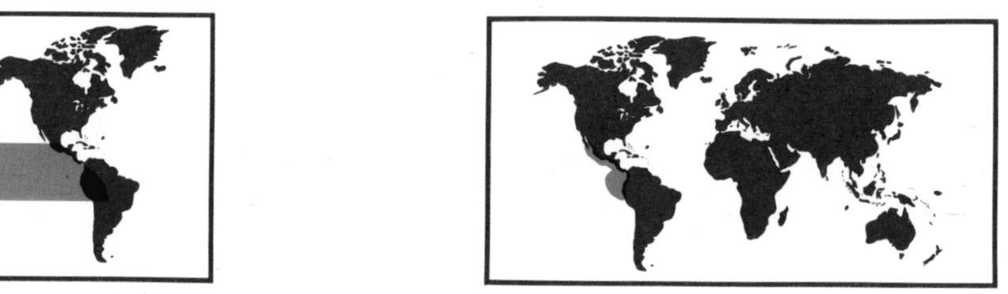

TURTLE POISONING

Most people are not aware of the poisoning caused by eating certain kinds of marine turtles, yet the cases that have been reported are severe enough to be taken seriously. As with fishes, most turtles are safely eaten by humans. However, some species from the vicinity of the Philippine Islands, Sri Lanka and Indonesia may become extremely poisonous to eat under certain conditions.

Family CHELONIIDAE

This family contains four genera comprising five species. The green sea turtle *Chelonia* usually inhabits water less than 25 m (82 feet) in depth and prefers areas sheltered by reefs, where it feeds on algae. It is also common in bays and lagoons. Occasionally, *Chelonia* will make its way into freshwater lakes. Green turtles are sometimes seen basking on reefs and beaches of islands uninhabited by man. They are primarily herbivorous, feeding upon the sea grasses *Cymodocea, Thalassia* and *Zostera,* and occasionally algae. When kept in captivity, they seem to show a preference for a diet of meat and fish. Green turtles nest between the latitudes 30 degrees north and 30 degrees south of the equator. They will migrate considerable distances, leaving their usual haunts to reach their breeding grounds. The nest site is usually selected on a beach having loose sand, within reach of the waves. When the site is chosen, the loose sand is brushed away with the front flippers. The entire nesting process requires about two hours, during which time a hole is dug and the eggs are laid.

The breeding season seems to be from July to November in Sri Lanka, but October to mid-February in Australia. The green sea turtle is considered one of the more valuable turtles for use as food.

The hawksbill turtle *Eretomochelys* is generally found close to land in tropical and subtropical oceans. Seldom does it enter lagoons. Like other turtles, it is predominantly a herbivore but may at times eat other foods. The breeding range is between 25 degrees north and south of the equator. Eggs are laid on sandy beaches in a manner similar to that used by *Chelonia.* As many as 115 eggs are laid at a time. The egg-laying season extends from November to February in some areas, but seems to take place during April to June in others. *Eretomochelys* is of commercial importance because of its overlapping scutes, which are used in the manufacture of jewelry and other products.

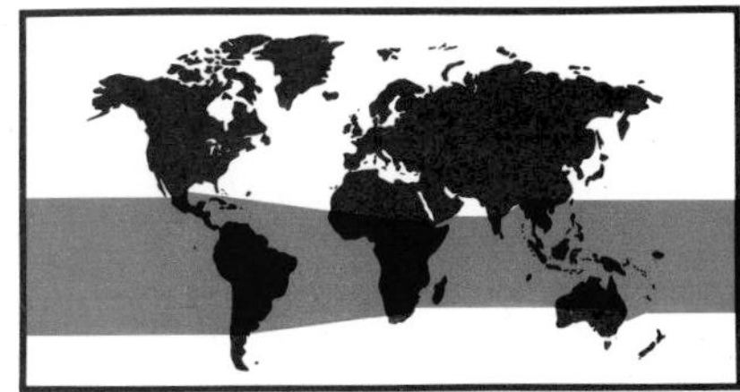

***Chelonia mydas* (Linnaeus)**
Green sea turtle

Length of shell 1.2 m (4 feet). Tropical and subtropical seas.

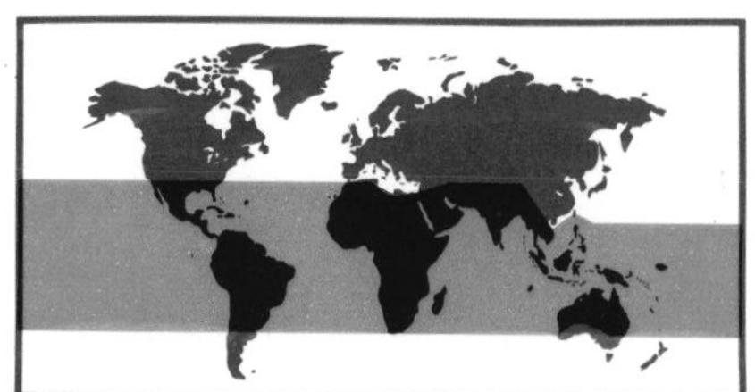

***Eretomochelys imbricata* (Linnaeus)**
Hawksbill turtle

Length of shell 85 cm (33 inches). Tropical and subtropical seas.

Family DERMOCHELIDAE

The leatherback turtle *Dermochelys* (the only genus in this family) is said to be the swiftest and largest of living chelonians, usually inhabiting relatively deep water near the edge of the Continental shelf. Newly-hatched leatherbacks head directly for the open ocean and do not return to shallow water until they are ready for egg laying. An adult may attain a weight of more than 780 kg (1720 pounds). Its food consists of algae, crustaceans and fishes. *Dermochelys* is believed to lay eggs three or four times a year, which takes place during May to June in Sri Lanka. The eggs are laid on sandy beaches at night. Often, several females will deposit their eggs in close proximity to each other.

***Dermochelys coriacea* (Linnaeus)**
Leatherback turtle

Length of shell 1.2 m (4 feet). Largely found in tropical waters, but sometimes taken in temperate waters.

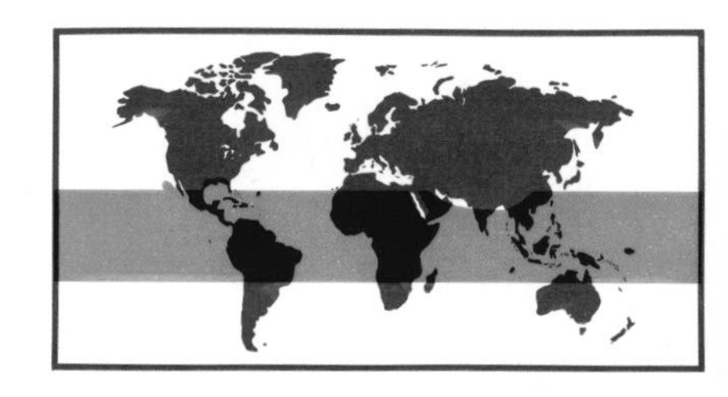

MARINE MAMMAL POISONING

Several kinds of marine mammals have been reported to cause human poisoning.

Family PHOCIDAE

The family of true seals consists of 13 genera and 18 species. Members are widely distributed throughout the coastal and oceanic waters of polar, temperate and tropical regions of the world.

Bearded seals (*Erignathus*) are solitary creatures, living alone on ice floes in the Arctic not far from land, except during the mating season. They derive their name from the beard-like tuft of stout white bristles growing down each side of the muzzle. Their food consists largely of crustaceans and mollusks; fishes are sometimes eaten but are apparently less desirable.

Bearded seals seek their food on the bottom of the sea and may dive to great depths. The single pup is born in late March. The adult male may attain a length of 2.7 m (9 feet) and a weight exceeding 360 kg (800 pounds). The flesh is said to be tough, coarse, and most appreciated by Eskimos when it is decomposed and frozen. The liver is toxic to eat, apparently because of excessive vitamin A content.

The Australian sea lion (*Neophoca*) is the largest of its kind, inhabiting rocky coastal areas along the southern shore of Australia. For the most part, these seals are non-migratory, usually remaining in the immediate environs of their birthplace. They are generally docile, except during the mating season, when they may become ill-tempered.

Their food consists mainly of penguins, which are available in abundance, and fishes. They also have the interesting habit of ingesting stones, apparently used as an aid in digesting food. The breeding season is from October to early December, during which time the community spends most of its time inshore. The harems are relatively small, consisting of one to four females to each male. The females give birth to a single pup. The male attains a maximum length of about 3.7 m (12 feet). The flesh of some species has been reported to be highly toxic.

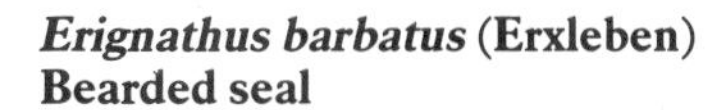

***Erignathus barbatus* (Erxleben)**
Bearded seal

Length up to 2.7 m (9 feet). Found in the Arctic, living at the edge of the ice.

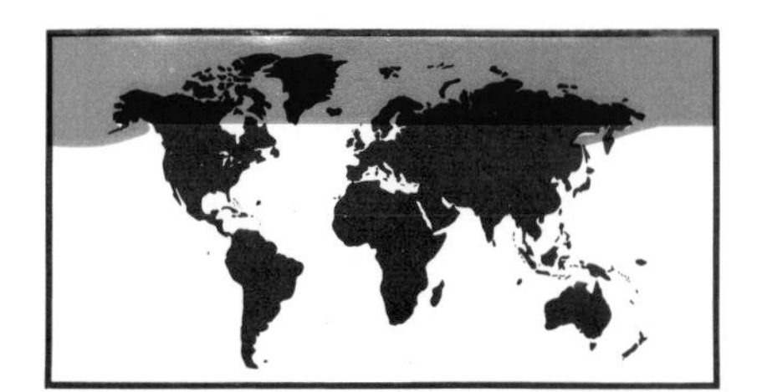

***Neophoca cinerea* (Peron)**
Australian sea lion

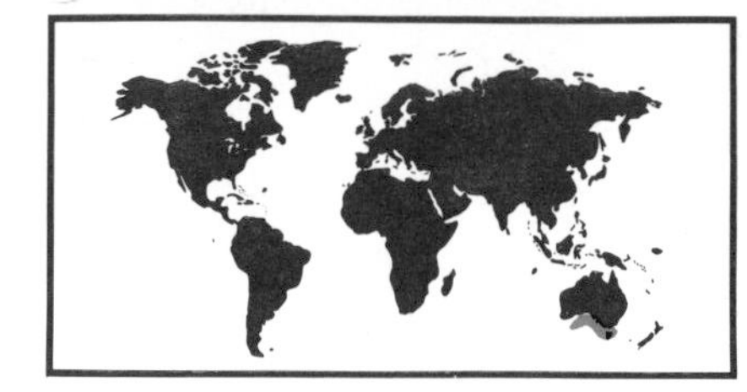

Length up to 2.4 m (8 feet). Confined to the coast of South Australia.

Family URSIDAE

The only marine carnivore toxic to man is the polar bear *Thalarctos maritimus* (Phipps). There is a single species. It is an inhabitant of the ice pack of the Arctic Ocean, where icebergs and broken pan ice are interspersed with stretches of open water. With the ever-shifting margin of the polar cap and the open ocean, the polar bear migrates according to the season. Seldom does it venture any great distance inland from the sea. It is a remarkable and powerful swimmer. Peary is said to have seen the tracks of a polar bear along the course of a lead covered with young ice, "about 200 miles [320 km] from land". Individual bears have actually been observed swimming in the open sea "more than 40 miles [65 km] from shore". Either the forepaws alone or all four feet may be used in swimming. The soles of the feet are covered with fur overshoes which lend protection and provide sure footing on the slippery ice. The polar bear enjoys a wide variety of foods: whale meat, birds, fish, tundra vegetation, or almost anything else that is available – but it is particularly fond of seals and young walruses. Many poisonings have resulted from eating the liver and kidneys of polar bears, which are reported to have lethal levels of vitamin A.

***Thalarctos maritimus* (Phipps)**
Polar bear

Length up to 2.5 m (8 feet); weight up to 750 kg (1650 pounds).

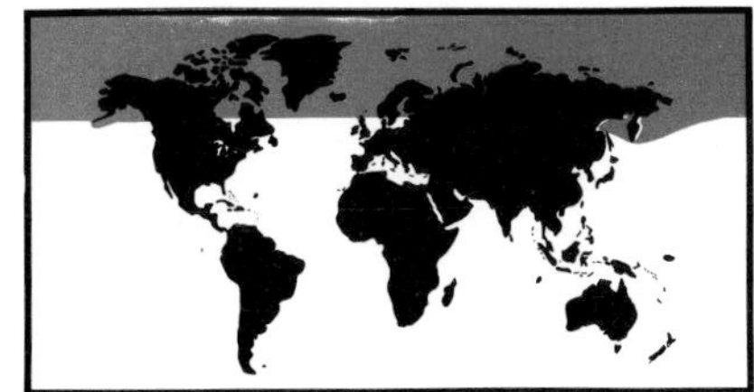

Order CETACEA

Mammals of the order Cetacea are characterized by their spindle-shaped body form. The head is long, often pointed, and joined directly to the body. Some species have a fleshy dorsal fin. The flippers or forelimbs are broad and paddle-like; the digits are embedded and have no claws. There are no hind limbs. The tail is long and ends in two broad, transverse, fleshy flukes notched in the midline. The teeth, when present, are all alike and lack enamel. The nostrils are on top of the head. The ear openings are minute. There are no skin glands except mammary and conjunctival glands. A thick layer of blubber under the skin affords insulation. The livers of some species may be toxic to eat.

Family BALAENOPTERIDAE

The fin back whales are the largest of living animals. The largest member of the group, the blue or sulphur-bottom whale, *Sibbaldus musculus* (Linnaeus), attains a length of 30 m (100 feet) and a weight of 112,500 kg (248,000 pounds). This family is composed of three genera and six species and occurs in all oceans. It is one of two families of baleen whales in which the embryonic teeth are replaced by baleen plates in the adult animal. Fin back whales are frequently called "rorquals", which refers to "a whale having folds or pleats". The rorquals are equipped with longitudinal furrows, usually 10 to 100 in number and 2.5 to 5 cm (1 to 2 inches) deep, which are present on the throat and chest. These furrows increase the capacity of the mouth when opened.

The members of this family are the fastest swimmers of the baleen whales, some attaining speeds of up to 48 km per hour (30 miles per hour). They usually travel singly or in pairs, but several hundred individuals may congregate when food is abundant. Their food consists largely of euphausiid shrimp, copepods, amphipods, and other zooplankton. Some species include fishes and penguins in their diet. The zooplankton are captured by gulping and swallowing, or skimming. When skimming, the whale swims through the zooplankton with its mouth open and its head above the surface of the water. When a mouthful of organisms has been filtered from the water by the baleen plates, the whale dives, closes its mouth and swallows the plankton. Rorquals breed and give birth in the warmer waters within their range. The large species give birth to a single calf every other year, but the smaller forms breed more frequently. Several members of this family are hunted commercially for their oil and meat. Whales are considered to be among the most healthy of all living mammals, since evidence of pathology is seldom observed. This family contains two genera with five species.

Balaenoptera borealis Lesson
Sei whale

Length up to 18 m (60 feet). Atlantic Ocean from the coast of Labrador southward to Baja California.

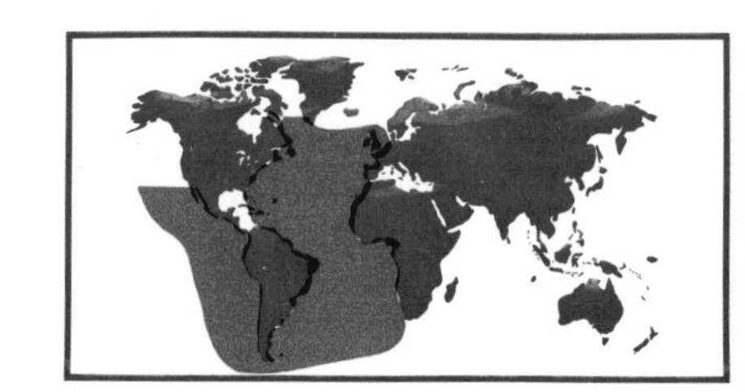

Family DELPHINIDAE

This family of dolphins and porpoises comprises 18 genera and about 62 species. They inhabit all the oceans and the estuaries of many large rivers; some species may ascend the rivers for great distances. Some species seem to prefer warm coastal waters and are never found in polar regions. The term "dolphin" generally refers to small cetaceans having a beak-like snout and slender streamlined body, whereas the term "porpoise" refers to small cetaceans having a blunt snout and rather stout, stocky body form.

Dolphins are among the most agile and swift swimmers. They are capable of speeds up to 25 knots and are frequently seen following ships and frolicking about the bow. They occur in schools from five to several hundred individuals. Migration is known to occur in some species. They utter a wide variety of underwater calls and noises. Cooperative behavior has often been observed in which one or more individuals will come to the aid of a fellow in distress or a female giving birth, pushing the young to the surface so that it can breathe. It is the opinion of those that have studied their behavior that they are highly intelligent animals. The killer whale *Orcinus* feeds mainly on cephalopods and fishes. The gestation period varies with the species between nine and 12 months, the calf being born under water. Some species are commercially hunted. *Neophocaena phocaenoides* ascends estuaries and rivers more than 1600 km (1000 miles) from the mouth. This species is usually seen singly or in pairs. They are sluggish in their movements and roll when rising to breathe. Birth of the young occurs around October. Their food consists of fishes, crustaceans and cuttlefish.

Neophocaena phocaenoides (Cuvier)
Southeast Asiatic porpoise, black finless porpoise

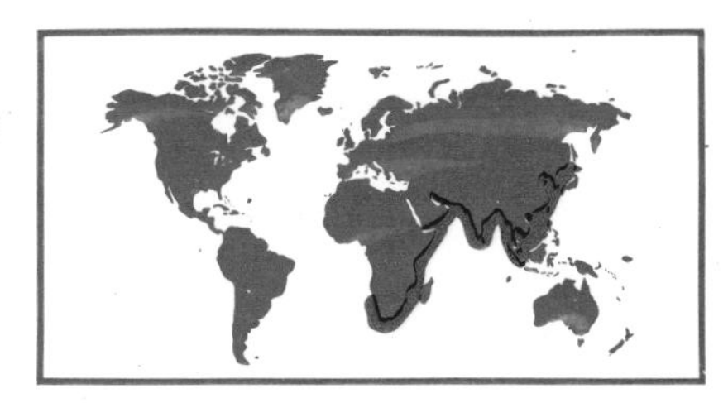

Length 1.5 m (5 feet). Frequents the coasts, estuaries, rivers and lakes of China, Japan, Borneo, Java, Sumatra, Pakistan, and the Indian Ocean to South Africa. It has been reported 1600 km (1000 miles) from the mouth of the Yangtze River and reaches Tungting Lake. Its liver may be toxic.

Family MONODONTIDAE

The family of white whales consists of only two genera, each having a single species: *Delphinapterus leucas*, the white whale, and *Monodon monoceros*, the narwhal. Both species are found in Arctic seas, but many ascend rivers. Ingestion of the flesh of the white whale has caused fatalities. There is no information available concerning the edibility of the narwhal. The white whale attains a length of about 5.5 m (18 feet) and a weight of about 900 kg (2000 pounds). The body shape resembles that of a member of the Delphinidae. The snout is blunt and there is no beak. There are no external grooves on the throat. White whales usually live in schools, sometimes consisting of more than 100 individuals. They migrate in response to the shifting pack ice and rigorous winter. The white whale can swim for hours at a speed of 9 km per hour (5.5 miles per hour) and remain under water for up to 15 minutes. Whales emit various sounds which are probably produced by emission of a stream of bubbles, rather than by the voicebox. They feed mainly on benthic organisms, cephalopods, crustaceans, and fishes. The gestation period is about 14 months, and the calf is about 1.5 m (5 feet) long at birth. White whales are of economic importance and are hunted mainly for their skins, which are sold as "porpoise leather". Livers are said to be toxic.

Delphinapterus leucas **(Pallas)**
White whale or beluga

Length 5.5 m (18 feet). Arctic and subarctic seas.

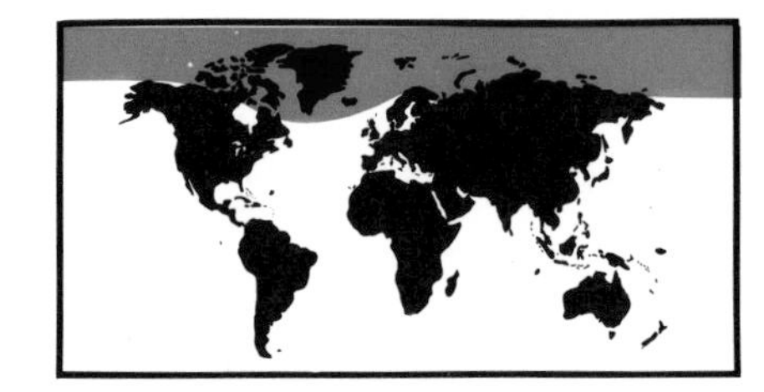

Family PHYSETERIDAE

The family of sperm whales consists of two genera and two species which inhabit all oceans. The sperm whale *Physeter catodon* attains a length of up to 20 m (65 feet) and can weigh more than 55 metric tons. The other member of the family, the pygmy sperm whale *Kogia breviceps*, attains a length of about 4 m (13 feet) and a weight of about 320 kg (700 pounds). The characteristic features of *Physeter* are its tremendous barrel-shaped head and underslung lower jaw. The sperm whale is said to be the only cetacean with a gullet large enough to swallow a man. It sometimes lifts its head out of the water to look and listen. When necessary, it can swim up to 12 knots per hour. They usually travel in groups up to 20 individuals, but large schools may number in the hundreds. The gestation period in *Physeter* is about 16 months. They feed on squid, cuttlefish, fishes, and elasmobranchs. *Physeter* is hunted primarily for the oil and spermaceti, used for making candles and ointments. Ambergris is a substance unique to the sperm whale and is believed to be formed from solid wastes coalescing around a matrix of indigestible matter. The meat of the sperm whale is usually discarded by pelagic whalers, but some of the Pacific coast stations freeze the meat as food for fur-bearing animals or treat it to yield oil. Livers are said to be toxic.

***Physeter catodon* (Linnaeus)**
Sperm whale

Length up to 20 m (65 feet). Polar, temperate and tropical seas.

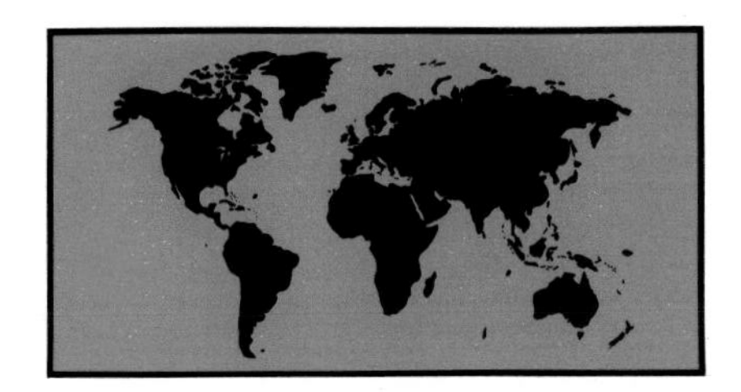

CHAPTER 6

ELECTROGENIC ANIMALS

Electricity is an important constituent in the metabolic activity of living things. The amount of current is normally so small that it can be detected only by extremely sensitive instruments which are used in laboratory procedures. In land animals, the air acts as a good insulator and these small discharges are difficult to detect at any distance from the body. Electric fishes possess a specialized organ system that discharges electricity through the water at surprisingly high voltages and is used to stun prey.

There are about 250 species of fish known to possess specialized electric organs capable of delivering a painful electric shock. Many other fish are able to use electric discharges as a form of detection mechanism during social interactions.

Of the several fishes that have electric organs, electric eels and catfishes live in fresh water, while stargazers and rays are marine. The most important marine members are the electric rays, which are found in all temperate and tropical oceans. Most electric rays are inclined to be sluggish, feeble swimmers, spending most of their time lying on the bottom buried in the mud or sand, generally staying in shallow depths.

The electric organs, which make up about one-sixth of the total body weight of the ray, are located on the front part of the body. The production of an electrical discharge or shock is believed to be a simple reflex action. A ray can deliver a series of shocks, which become weaker, until it is finally exhausted. After a period of time, the ray is again prepared to produce electrical discharges. The voltage power is different with each individual species, but may range from 8 to 220 volts. One such species is the lesser electric ray, *Narcine brasiliensis* (Olfers), which occurs in inshore waters of the western Atlantic from Brazil to Florida and Texas, and south to Argentina.

The average adult *Narcine* produces a discharge of about 14 to 37 volts. Other species, e.g. *Diplobatis* and *Torpedo*, are said to produce discharges of up to 200 volts. The Greeks knew the torpedo rays and referred to them as the "narke", a term from which derives the words narcotic and narcosis. Many physicians of ancient Rome used these torpedo rays to shock their patients.

***Torpedo marmorata**, marbled crampfish or torpedo ray*

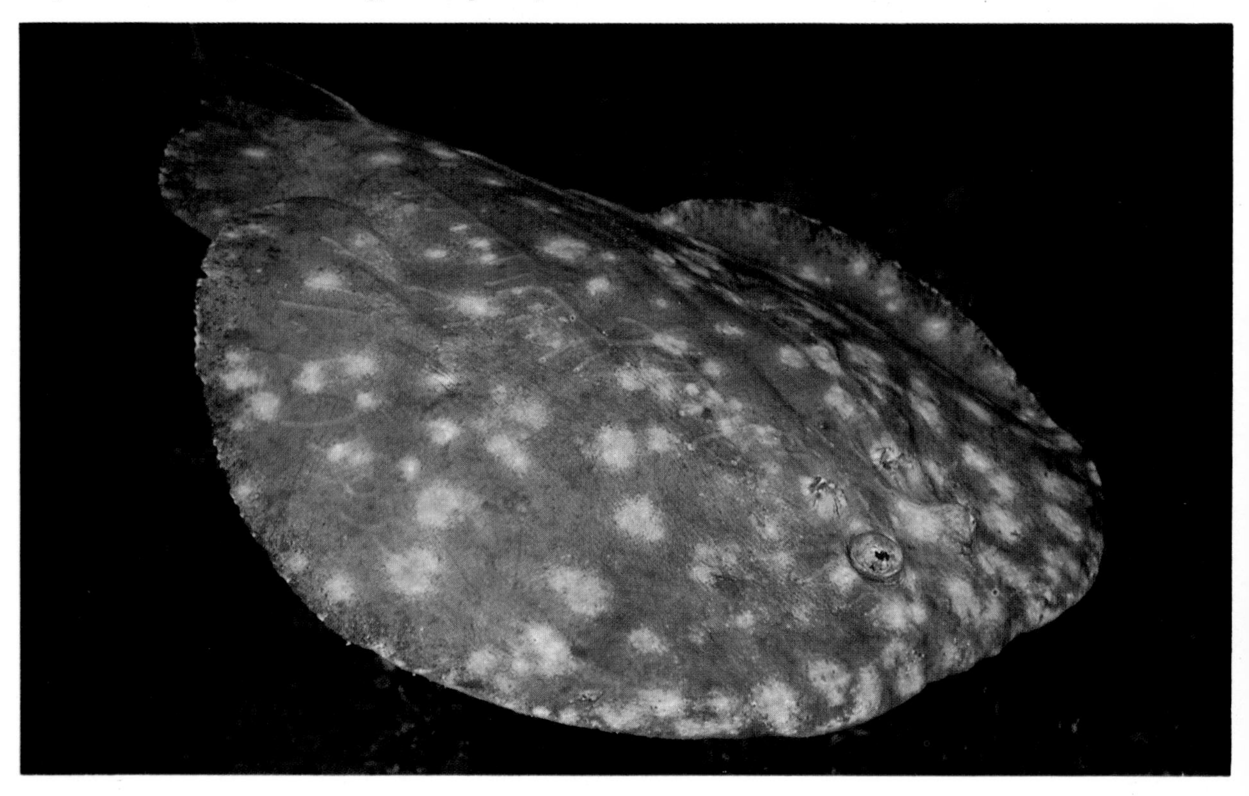

ELECTRIC RAYS

Family TORPEDINIDAE

This family contains 10 genera with about 38 species. They are marine electric rays with powerful electric organs, one on either side of the anterior part of the disk between the anterior extension of the pectoral and the head, extending forward to about the level of the eye and posteriorly past the gill region to the vicinity of the pectoral girdle.

The electric rays are characterized by their disk, which is subcircular to elongate, fleshier toward its margins and thicker than in most other disk-shaped batoids, and by their softer body. The tail is with or without lateral folds, sharply marked off from the body portion and broader in the base than in most other batoids. The attachment of the anterior portion of the pectoral extends forward to or beyond the level of the eyes. There are one or two well-developed fins. The caudal fin is well developed. The skin is soft and entirely naked. These are some of the essential characteristics which separate this group from other members of the ray family, although most possess weak electric organs at the base of the tail.

Although most species seem to prefer shallow waters, some members are found at moderately greater depths. Their food consists of crustaceans, mollusks, worms, other vertebrates and fishes. Electric rays are inhabitants of temperate, subtropical and tropical latitudes. They attain a length of about 2 m (6.5 feet) and a weight of over 91 kg (200 pounds). Development is ovoviviparous. A rested ray is said to be able to produce an electrical discharge sufficient to knock down and temporarily disable an adult man.

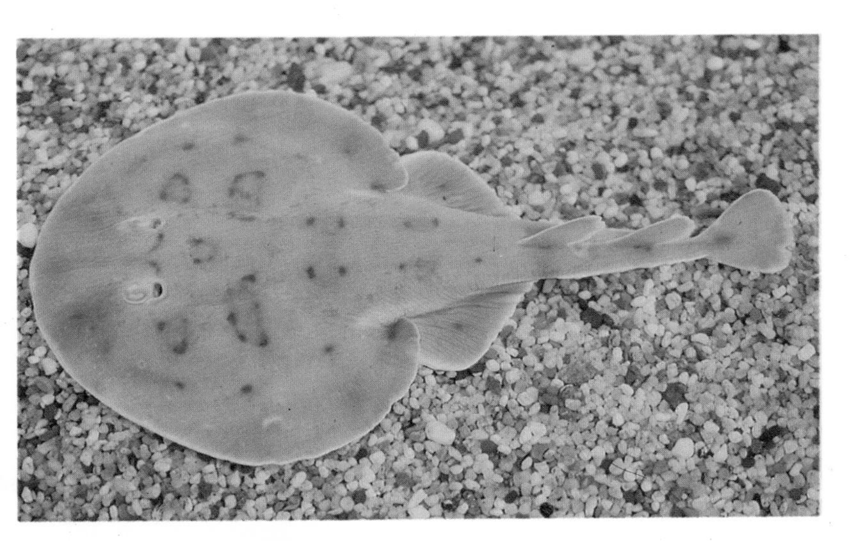

***Narcine brasiliensis* (Olfers)**
Lesser electric ray

DESCRIPTION Length about 28 cm (11 inches). Usually has a deep groove around the mouth and lips with long and strongly protractile jaws. Can discharge between 14 and 37 volts.

HABITAT Usually found in the shallow waters of the western Atlantic, from Florida and Texas south to Argentina.

OTHER POINTS Contact with the skin surface of the ray can produce a severe electric shock. The polarity of the electric organ is positive on the dorsal side and negative on the underside.

***Torpedo marmorata* (Risso)**
Marbled crampfish or torpedo ray

DESCRIPTION Length 70 cm (27 inches); yellowish coloration covered by many dark brown spots, which blend together to form a marbled pattern.

HABITAT Primarily inhabits sandy floors in the Mediterranean Sea.

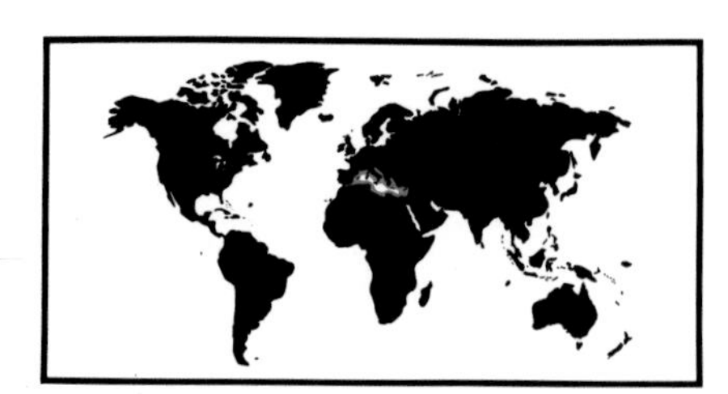

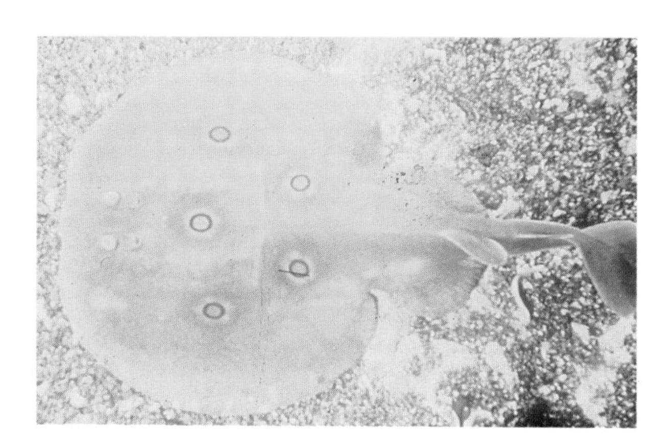

***Torpedo torpedo* (Linnaeus)**
Torpedo or crampfish

DESCRIPTION Length 60 cm (24 inches); brownish back with obvious eye spots. Like all electric rays, it is viviparous. This is the electric ray which is most often encountered.

HABITAT Shallow water to 50 m (165 feet), in the coastal Mediterranean and the eastern Atlantic from the Bay of Biscay to Angola.

OTHER POINTS All torpedo rays are capable of inflicting a severe shock of up to 200 volts.

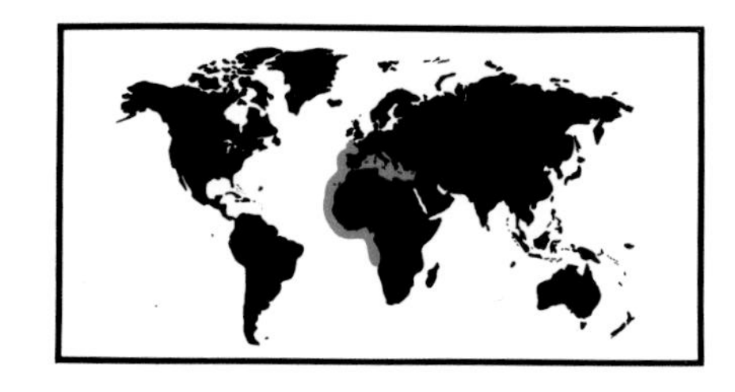

STARGAZERS

Family URANOSCOPIDAE

Another interesting group of electric fishes includes the stargazers of the family Uranoscopidae, which contains eight genera and about 25 species. The stargazers are a group of small, carnivorous and bottom-dwelling marine fishes. They are characterized by having a large cuboid head, an almost vertical mouth with fringed lips, and an elongate, conical, and subcompressed body. The eyes are small and anteriorly situated on the flat upper surface of the head. The family contains less than a dozen species, most of which are about 40 cm (16 inches) or less in length. Despite the small number of species, representatives of the family are distributed throughout the Mediterranean Sea and warmer parts of the Pacific, Atlantic and Indian Oceans. Their electrogenic organs are said to be modified eye muscles. When at rest, *Astroscopus* discharges only about 90 millivolts, but is capable of releasing up to 50 volts when necessary. Stargazers are also of medical importance by virtue of their venomous cleithral or shoulder spines. Since uranoscopids spend a considerable portion of their time buried in the sand or mud with only their eyes and a portion of their mouth protruding, they present a possible menace to intruders.

Astroscopus y-graecum **(Cuvier and Valenciennes)**
Southern stargazer

DESCRIPTION Length 38 cm (15 inches). Typical of uranoscopids, the eyes are on the top of the skull and the body is long, almost funnel-shaped.

HABITAT South Atlantic coast from Cape Hatteras to the Caribbean Sea. Usually lies quietly in the sand.

OTHER POINTS Can discharge up to 50 volts. The electric organs are located in a specialized pouch behind the eyes.

CHAPTER 7

ADVANCED MEDICAL TREATMENT

While standard emergency medical techniques usually suffice in managing marine animal-induced injuries and poisonings, there are a number of unique features of therapy related to special considerations of toxins and the bacteriological environment. New developments in diagnosis and therapy occur regularly, so the astute medical professional should pay close attention to recent literature. The advice offered in this section reflects our current understanding of marine-acquired injuries and illnesses.

CIGUATERA POISONING

If possible, a piece of the fish should be obtained for ciguatera toxin bioassay. Therapy is supportive and based upon symptoms. Gastric lavage or syrup of ipecac-induced emesis, followed by the administration of a slurry of activated charcoal (100 gm) in sorbitol may be of limited value if performed within three hours of ingestion. It has been suggested that the administration of magnesium-containing cathartics may augment a calcium channel blockade, so these should be avoided. Nausea and vomiting may be controlled with an antiemetic (prochlorperazine 2.5 mg IV or hydroxyzine 25 mg IM).

The most worrisome systemic problem is hypotension, which may require the administration of intravenous crystalloid for volume replacement and, rarely, pressor drugs such as dopamine or dobutamine. Some investigators recommend the administration of calcium gluconate (1-3 gm IV over 24 hours) for the management of hypotension and myocardial failure, although clinical hypocalcemia is not a diagnostic standard. Infusion of mannitol has been reported to alleviate the severe symptoms of cardiovascular failure. Bradyarrhythmias that lead to cardiac insufficiency and hypotension generally respond well to atropine (0.5 mg IV up to 2.0 mg). Cool showers or the administration of hydroxyzine (25 mg po q 6-8 h) have been recommended to relieve pruritus. Moderate headache may be extraordinarily responsive to acetaminophen. Other drugs that have been recommended at one time or another include edrophonium, neostigmine, corticosteroids, pralidoxime chloride, ascorbic acid, pyridoxine chloride (vitamin B6) and salicylic acid-colchicine-vitamin B complex. Amitriptyline (25 mg po bid) has been reported to be effective in the amelioration of pruritus and dysesthesias, as has protamide.

During recovery from ciguatera poisoning, the victim should exclude the following from the diet: fish (fresh or preserved), fish sauces, shellfish, shellfish sauces, alcoholic beverages, nuts and nut oils. All of these may result in an exacerbation of the ciguatera syndrome.

COELENTERATE STINGS

Recommended first aid measures do not interfere with any subsequent therapeutic measures likely to be employed. Further treatment should be directed toward three objectives: neutralizing the effects of the poison; relieving pain; and controlling systemic effects. Most coelenterate stings are relatively mild and require minimal care. The application to the skin injury of meat tenderizer (papain), vinegar, local anesthetic creams, ointments, lotions or aerosols and antihistaminics has been used with varying degrees of success. Occasional paradoxical reactions have been noted with benzocaine, which may be a skin sensitizer.

The physician should always be prepared to manage hypotension with prompt intravenous administration of crystalloid fluids. This must be done in concert with detoxification of any nematocysts that are still attached to the victim. Bronchospasm may be managed as an allergic component. Supplemental oxygen should be administered by face mask. All elderly victims should be evaluated with an electrocardiogram and be observed for arrhythmias. Urine should be examined for the presence of hemoglobinuria. Pain may be controlled by treating the dermatitis. However, if pain is excruciating and there is no contraindication, then the administration of narcotics is often indicated.

Serious stings from *Chironex* may require prompt antivenin therapy so as to neutralize the venom as rapidly as possible. The Commonwealth Serum Laboratories, Melbourne, Australia, produce a highly effective *Chironex* antivenin. The dose can be administered either intravenously or else intramuscularly. Use of the intramuscular route is justified if the person giving the antivenin is not trained in the technique of intravenous therapy. Box jellyfish antivenin should be stored in a refrigerator at 2 to 10°C (35 to 50°F) and must not be frozen. Although it is useful in treating other types of coelenterate stings, it is rarely required. Although the antivenin is prepared by hyperimmunizing sheep, the risk of anaphylaxis or serum sickness should be assumed to be the same as for equine hyperimmune globulin preparations. Hypersensitivity reactions can usually be controlled with epinephrine, corticosteroids and antihistaminics. Hypotension may be managed with intravenous volume expansion and prèssor agents if necessary. It cannot be over-emphasized that timely administration of antivenin can be lifesaving. In addition, early administration of antivenin may markedly reduce pain and decrease subsequent skin scarring.

CONE SHELL POISONING

The patient should rest as much as possible and be moved to a hospital as soon as is convenient, with the affected extremity kept in a dependent position. Respiratory distress may develop, requiring advanced life support with artificial ventilation until the patient can be placed on a mechanical ventilator. When the patient arrives at a hospital, intensive care may be required. Respiratory stimulants are of little value because of the neuromuscular blocking action of the cone shell venom. In cases of severe persistent hypotension, naloxone 2 to 4 mg may be administered to attempt to block the beta-endorphin vasodepressor response. The etiology of coagulopathy following cone shell envenomation has not been well-delineated, because of the infrequency of occurrence, and therapy varies with local custom.

CORAL CUTS

Once irrigated, the wound may be managed in a number of fashions, In no case should a coral cut ever be sutured tightly closed, as infection will surely follow. A preferred technique is to apply sterile wet-to-dry dressings utilizing normal saline or dilute antiseptics. Dressings are changed daily. The addition of non-toxic, topical and water-soluble powdered antibiotics is of no proven benefit. An alternative therapy is to cover the wound with a nonadherent dressing over an antiseptic ointment. Despite the best efforts at primary irrigation and decontamination, the wound may heal slowly, with moderate to severe soft tissue inflammation and ulcer formation. All devitalized tissue should be debrided regularly by sharp dissection. This should be continued until a bed of healthy granulation tissue is seen to form. Wounds that appear infected should be cultured and treated with antibiotics.

MARINE WORM TRAUMA

If the inflammatory reaction becomes severe, the victim may benefit from topical or systemic corticosteroids. Antibiotics may be required to treat secondary infection.

OCTOPUS BITES

There is no effective antivenin for the bite of the blue-ringed octopus. The toxin contains at least one fraction identical to tetrodotoxin, which blocks peripheral nerve conduction by interfering with sodium conductance in excitable membranes. This paralytic agent produces neuromuscular blockade, notably of the phrenic nerve supply to the diaphragm, without any apparent direct cardiotoxicity.

Treatment is based on the symptoms and is supportive. Prompt mechanical respiratory assistance has by far the greatest influence on outcome. Respiratory demise should be anticipated early and the rescuer prepared to provide artificial ventilation, including endotracheal intubation and the application of a mechanical ventilator. The duration of intense clinical venom effect is four to ten hours, after which the victim who has not suffered an episode of significant hypoxia will show rapid signs of improvement. Complete recovery may require two to four days.

Treatment of the bite is open to controversy. Some doctors recommend wide circular excision of the bite wound down to the deep fasica, with primary closure or immediate full thickness free skin grafts, while others advocate observation and a non-surgical approach. Because the local tissue reaction is not a significant cause of morbidity, excision is putatively recommended to remove any sequestered venom.

PARALYTIC SHELLFISH POISONING

If the victim is treated within the first few hours after ingestion, the stomach should be emptied with gastric lavage or emesis induced by syrup of ipecac. Lavage with alkaline fluids, such as a solution of 2% sodium bicarbonate or ordinary baking soda, is said to be of value, since the poison is acid-stable. Following gastric emptying and alkaline irrigation, the victim should be given activated charcoal (50 to 100 gm) in a sorbitol-based solution. Magnesium-based cathartics should be avoided.

The greatest danger is respiratory paralysis, which may require advanced life support. Oxygen should be administered and mechanical ventilation introduced where appropriate. Any victim who may have ingested saxitoxin should be rushed to an emergency facility for observation and therapy. The development of antibodies directed against saxitoxin may soon allow more specific treatment.

SCOMBROID POISONING

Therapy is directed at reversing the histamine effect. Minor intoxications can be treated with intravenous diphenhydramine or hydroxyzine (also an antiemetic), supported as necessary with subcutaneous epinephrine, inhaled bronchodilators, intravenous steroids and pressor agents. Anaphylaxis is uncommon and should be managed in standard fashion. Persons who are being medicated with isonicotinic acid hydrazide may have severe reactions due to INH blockade of gastrointestinal tract histaminase. Monoamine oxidase (MAO) inhibitors pose a theoretical hazard. If large quantities of the tainted fish have been consumed within an hour of presentation of the patient to an emergency facility, it may be of value to empty the stomach and administer activated charcoal with a cathartic (if the patient does not already have diarrhea). The histamine-2 antagonists cimetidine (300 mg IV) or ranitidine (50 mg IV) may be rapidly efficacious in alleviating symptoms in patients who do not respond to diphenhydramine. These should be used cautiously when combined with a histamine-1 blocker to avoid hypotension.

Nausea and vomiting are usually controlled by the diphenhydramine, but occasionally require the addition of a specific antiemetic such as prochlorperazine (2.5 mg IV). If the allergic and gastroenteric components are severe, the cumulative effect may induce hypotension, which will require the administration of intravenous crystalloid solutions and, rarely, pressor agents.

SEA SNAKE BITES

Sea snake envenomation is a medical emergency requiring immediate action. Delay in obtaining proper medical treatment can lead to consequences far more tragic than might be incurred with an ordinary traumatic injury. The first step is to determine if envenomation has occured. With any evidence of envenomation, polyvalent sea snake antivenin (Commonwealth Serum Laboratories, Melbourne, Australia) prepared from the venoms of *Enhydrina schistosa* and *Notechis scutatus* should be administered after appropriate skin testing for equine serum hypersensitivity. Conjunctival testing for hypersensitivity is no longer recommended. If this antivenin is not available, *Notechis scutatus* antivenin should be used. Monovalent *E. schistosa* antivenin is not yet widely available. Sea snake antivenin is specific and absolutely indicated in cases of envenomation. Supportive measures, while critical in management, are no substitute.

The administration of antivenin should begin as soon as possible, and is most effective if initiated within eight hours of the bite. The minimum effective adult dosage is one ampule (1000 units), which neutralizes 10 mg of *E. schistosa* venom. Three thousand to 10,000 units may be required, depending upon the severity of the envenomation. The proper administration of antivenin is described clearly on the antivenin package insert. The antivenin should be administered intravenously if possible, as the intramuscular route is less controlled and may be less effective. There is no reason to inject the antivenin into or near the bite site, and it should never be injected into a finger or toe.

If the clinical situation permits, a skin test should be performed for sensitivity to horse serum. This test should only be done after the decision to administer antivenin has been made and not to determine whether or not antivenin is to be given. The skin test is performed with an intradermal injection of 0.02 ml of a 1:10 dilution of horse serum test material in saline, with 0.02 ml of saline in the opposite extremity as a control. Erythema and pseudopodia will be present in 15 to 30 minutes in a positive response. Because antivenin contains many times the protein content of horse serum used for skin testing, the use of antivenin for skin testing may increase the risk for an anaphylactic reaction. If the skin test is positive, the antivenin should be diluted in sterile water or saline to a 1:100 concentration for administration.

If hypersensitivity reactions occur, they can usually be controlled with corticosteroids, epinephrine, and antihistamines. In cases of severe hypersensitivity, an intravenous epinephrine infusion may be necessary. Emergency desensitization with graded doses of antivenin may be performed. The purification of antivenin products is an area of continuing research that may soon yield antibody fragments of high specificity and low allergic potential.

Supportive measures such as blood transfusions, vasoregulatory support with pressor agents, antibiotics, oxygen and mechanical ventilation may be required. More severe reactions may be anticipated in the very young, elderly and infirm.

SEA URCHIN WOUNDS

If there is any doubt as to whether a spine is present, soft tissue density radiographic techniques may be employed. Some fragmented spines will be absorbed in a few days and disappear. All thick calcium carbonate spines should be removed to avoid secondary infection or the development of a troublesome foreign body encaseation granuloma or dermoid inclusion cyst.

Sea urchin spines which penetrate into a joint or directly involve a neurovascular structure should be surgically removed. This should be performed in a properly equipped operating theater using an operating microscope. If the spine has entered an interphalangeal joint, the finger should be splinted acutely until spine exaction is performed, to limit fragmentation and further penetration. The use of organic solvents to attempt to dissolve spines is ineffective and potentially tissue toxic. External percussion to achieve fragmentation may prove disastrous. Because sea urchin wounds frequently become infected, antibiotic therapy may be required.

Pedicellariae envenomations may produce severe reactions because they involve venom more potent than that found in the spines. After the pedicellariae are removed, the wound should be bathed with acetic acid 5% (vinegar) or isopropyl alcohol 40 to 70% if the hot water technique is not effective. Bronchospasm and other allergic reactions can usually be controlled with epinephrine and antihistaminics. Unfortunately, there are no specific antivenins available for sea urchin stings.

SHARK BITES

Continue to initiate aggressive therapy to remedy hypovolemic shock. Type and cross-match the victim for whole blood or packed red blood cells immediately to allow expeditious transfusion if required. Monitor urine output.

Examine the victim thoroughly for evidence of cervical, intrathoracic or intraabdominal injuries. If there is a possibility of orthopedic injury, radiological examination is required. Any small child who has sustained a scalp laceration associated with a shark bite should undergo skull radiography for a possible occult fracture. All significant wounds should be explored and debrided in a proper operating suite to remove foreign material, including shark teeth or pieces of neoprene wet suit, and devitalized tissue, while providing access for copious irrigation. All wounds should be closed loosely around drains using a minimum of deep sutures. Acute plastic surgical repairs should be conservative in nature during the initial period; reconstructive surgery may be required at a later date. Tetanus prophylaxis is mandatory. The victim should receive prophylactic antibiotics; acceptable agents include imipenem-cilastatin, gentamicin, amikacin, tobramycin, chloramphenicol, trimethoprim-sulfamethoxazole, or a third-generation cephalosporin (e.g., cefoperazone, cefotaxime, or ceftazidime). Penicillin and first-generation cephalosporins are not acceptable alternatives. After an infection is recognized, culture the wounds for aerobes and anaerobes by inserting sterile swabs deeply into purulent lesions.

SPONGE STINGS

Although corticosteroid lotions may help to relieve secondary inflammation, they are of little value in the initial decontamination. Following treatment with acetic acid, a mild emollient cream or steroid preparation may be applied to the skin. If the allergic element is severe, particularly if there is weeping, crusting, and blister formation, systemic corticosteroids should be used. Severe itching may be controlled with an antihistamine.

STARFISH WOUNDS

Spines are sometimes embedded and may have to be surgically removed, in order to avoid formation of a granulomatous lesion similar to that from a sea urchin puncture. If the spine is embedded, infiltration of the wound with 1 to 2% lidocaine without epinephrine may be required for pain control. Because of the stout nature of the spines, it is rare to retain a fragment. The puncture should be irrigated and explored to remove all foreign material. If there is doubt as to the presence of a foreign body, a soft tissue radiograph will often identify any fractured spine. Contact dermatitis is best treated with topical solutions, such as calamine with 0.5% menthol, or corticosteroid preparations.

VERTEBRATE FISH STINGS

Heat may have an attenuating effect on the heat-labile components of the venom, as is seen with stingray venom *in vitro*. The addition of solvents, potassium permanganate, magnesium sulphate or epsom salts to the water is of no particular benefit, and may be harmful if the additive is tissue toxic. Recurrent pain that develops after an interval of two to three hours may respond to a repeat hot water treatment. Intravenous calcium gluconate injections sometimes help to alleviate muscle spasms. Infiltration of the wound area with 0.5 to 2% lidocaine without epinephrine has been used with good results. If local measures or regional nerve blocks are unsatisfactory, narcotic analgesia will generally be efficacious. During the soaking procedure, all obvious pieces of spine and sheath fragments should be gently removed from the wound. Following the soaking procedure, debridement and further cleansing of the wound are mandatory. If foreign bodies have penetrated deeply into the sole of the foot, surgical exploration should be performed in the operating room with magnification. Wounds which do not require immediate repair for cosmetic reasons should be left open or closed loosely around packing or generous drainage. The injured area should be covered with an antiseptic and sterile dressing. After a period of four to five days, secondary closure may be performed.

Because of the propensity for wound infections, antibiotic therapy should be initiated in a manner analogous to that for shark bites. If delay has resulted and an infection is already present, the wound should be cultured for aerobes and anaerobes prior to administration of antibiotics. Tetanus prophylaxis is standard.

Hypotension which results from the action of stingray venom on the cardiovascular system requires immediate and vigorous therapy. Treatment should involve fluid replacement and the maintenance of vascular tone.

Indo-Pacific stonefish (*Synaceja* species) stings are extremely painful and can be lethal. Fortunately, the Commonwealth Serum Laboratories, Melbourne, Australia, produce an effective antivenin. The antivenin also should be considered for the treatment of severe systemic reactions following envenomation by other scorpionfish species. The antivenin is prepared by hyperimmunizing horses with stonefish venom. One unit of stonefish antivenin neutralizes 0.1 mg of stonefish venom (1000 units will neutralize 10 mg of venom). Each stonefish spine produces about 5 to 10 mg of venom. The initial dose of antivenin will depend upon the number of puncture wounds received by the victim. The antivenin should be stored protected from light at 2 to 8°C (35 to 46°F) and should not be frozen. It should be diluted in 50 to 100 ml of normal saline and administered slowly intravenously. The intramuscular route is not recommended, as absorption may be erratic. The physician who administers antivenin should be prepared to manage an anaphylactic reaction.

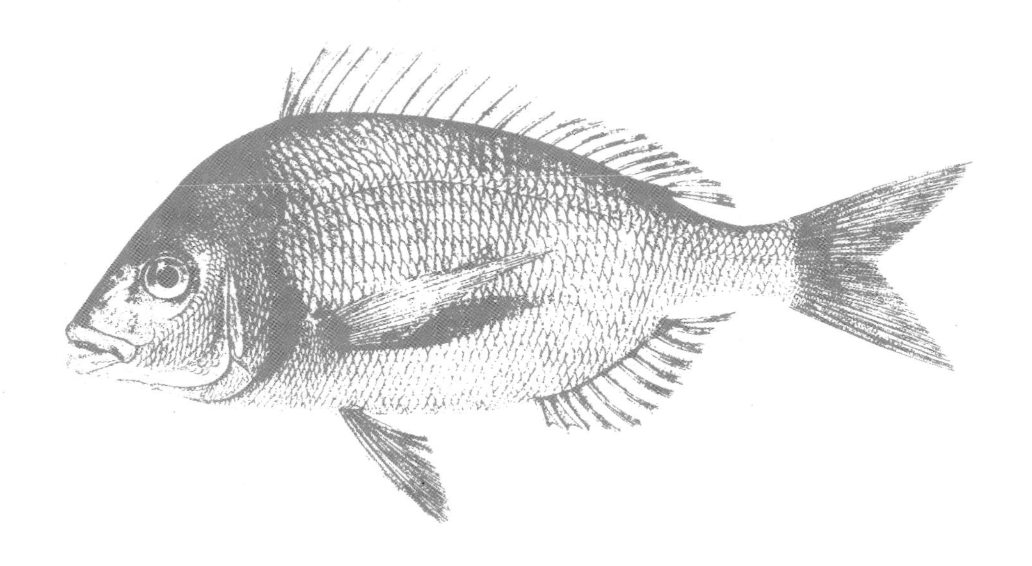

BIBLIOGRAPHY

1 Anderson, D.M., White, A.W., and D.G. Baden, [Eds.] 1985, *Toxic Dinoflagellates*. Elsevier Science Publishing Co., Inc., New York, (561 p).

2 Auerbach, P.S., and E.C. Geehr, [Eds.] 1989, *Management of Wilderness and Environmental Emergencies*. C.V. Mosby Co., St. Louis, (1068 p).

3 Baldridge, H.D., 1974, *Shark Attack*. Berkley Publishing Corp., New York, (263 p).

4 Brown, T.W., 1973, *Sharks–The Search for a Repellent*. Angus and Robertson Publishers, Sydney, Australia, (134 p).

5 Caras, R.A., 1964, *Dangerous to Man: Wild Animals–A Definitive Study of Their Reputed Dangers to Man*. Chilton Books, Philadelphia, (433 p., illust).

6 1974, *Venomous Animals of the World*. Prentice-Hall, Inc. Englewood Cliffs, (362 p).

7 Castro, J.I., 1983, *The Sharks of North American Waters*. Texas A&M, University Press, College Station, (180 p).

8 Coppleson, V., 1982, *Shark Attack*. Angus and Robertson Publishers, Sydney, Australia, (199 p).

9 Covacevich, J., Davie, P., and J. Pearn, [Eds.] 1987, *Toxic Plants and Animals: A Guide for Australia*. Queensland Museum, S. Brisbane, Australia, (501 p).

10 Davies, D., 1964, *About Sharks and Shark Attack*. Shuter and Shooter, Pietermaritzburg, S. Africa, (237 p).

11 Gilbert, P.W., [Ed.] 1963, *Sharks and Survival*. D.C. Heath and Company, Boston, (578 p).

12 Gopalakrishnakkone, P. and C.K. Tan, [Eds.] 1987, *Progress in Venom and Toxin Research*. National University of Singapore, Republic of Singapore, (747 p).

13 Halstead, B.W., 1965, *Poisonous and Venomous Marine Animals of the World*. U.S. Government Printing Office, Washington, D.C., (**1**:1-1994, illust).

14 1967, *Ibid.* (**2**:1-1070, illust).

15 1970, *Ibid.* (**3**:1-1006, illust).

16 1978, *Poisonous and Venomous Marine Animals of the World*, (rev. ed.) Darwin Press, Princeton, (1043 p., illust).

17 Halstead, B.W., 1980, *Dangerous Marine Animals*. Cornell Maritime Press, Centerville, (208 p., illust).

18 1988, *Poisonous and Venomous Marine Animals of the World*, (2nd rev. ed.) Darwin Press, Princeton, (1168 p., illust).

19 Hashimoto, Y., 1979, *Marine Toxins and Other Bioactive Marine Metabolites*. Japan Scientific Societies Press, Tokyo, Japan, (369 p).

20 Keeler, R.F. and A.T. Tu, [Eds.] 1983, *Handbook of Natural Toxins: Plant and Fungal Toxins*. Marcel Dekker, Inc., New York, (**1**:1-934).

21 Mandojana, R.M., [Ed.] 1987, *Aquatic Dermatology*. J.B. Lippincott Co., Philadelphia, (5(3):**1**-173).

22 Ohsaka, A., Hayashi, K., and Y. Sawai, [Eds.] 1976, *Animal, Plant and Microbial Toxins: Biochemistry*. Plenum Press, New York, (**1**:1-555).

23 1976, *Animal, Plant and Microbial Toxins: Chemistry, Pharmacology and Immunology*. Plenum Press, New York, (**2**:1-562).

24 Ragelis, E.P., [Ed.] 1984, *Seafood Toxins*. American Chemical Society. Washington, D.C. (460 p).

25 Russell, F.E., 1971, *Poisonous Marine Animals*. T.F.H. Publishing, Inc., Neptune City, (176 p., illust).

26 1980, *Snake Venom Poisoning*. J.B. Lippincott Co., Philadelphia, (562 p).

27 Russell, F.E., Gonzalez, H., Dobson, S.B., and J.A. Coats, 1984, *Bibliography of Venomous and Poisonous Marine Animals and Their Toxins*. Office of Naval Research, U.S. Government Printing Office, Washington, D.C. (416 p).

28 Smitinand, T. and W.R. Scheible, 1966, *Edible and Poisonous Plants and Animals of Thailand*. Joint Thai-U.S. Military Research and Development Center, Bangkok, Thailand, (250 p).

29 Springer, V.G., and J.P. Gold, 1989, *Sharks in Question: The Smithsonian Answer Book*. Smithsonian Institution Press, Washington, D.C. (187 p).

30 Strauss, R.H., [Ed.] 1976, *Diving Medicine*. Grune and Stratton, New York, (420 p).

31 Sutherland, S.K., 1983, *Australian Animal Toxins: The Creatures, Their Toxins and Care of the Poisoned Patient*. Oxford University Press, Melbourne, Australia, (527 p).

32 Tu, A.T., 1977, *Venoms: Chemistry and Molecular Biology*. John Wiley & Sons, Inc., New York, (560 p).

33 Tu, A.T., [Ed.] 1984, *Handbook of Natural Toxins: Insect Poisons, Allergens, and Other Invertebrate Venoms*. Marcel Dekker, Inc., New York, (**2**:1-732).

34 1988, *Handbook of Natural Toxins: Marine Toxins and Venoms*. Marcel Dekker, Inc., New York, (**3**:1-587).

35 Zahuranec, B.J., [Ed.] 1983, *Shark Repellents from the Sea: New Perspectives*. Westview Press, Inc., Boulder, (210 p).

For information on continuing research in the field of toxicology refer to the journal *TOXICON*, Pergamon Press, Oxford, England.

ACKNOWLEDGEMENTS

The Publishers thank the following institutions, agencies, photographers and other individuals for providing the illustrations that appear in this volume:

World Life Research Institute, Colton, California: pages 9 (Jackie Bond), 10 top, 10 bottom (Oceanographic Research Institute, Durban, South Africa), 12 bottom (R.H. Johnson and D.R. Nelson), 14 bottom (C. Roessler), 16 top (A. Gidding), 17 bottom (Ron Taylor Film Productions, Sydney, Australia), 18 top, 18 bottom (Ron Taylor Film Productions, Sydney, Australia), 20 top (R. and V. Taylor), 20 bottom (Al Gidding), 21 top, 23 top, 27 (D. Woodward), 28, 29 top (P. Auerbach), 29 insert, 33 (D. Ludwig), 34 top row (G. Webb), 36 (State of Alaska Department of Fish and Game, courtesy R. Wallen), 41 left (G. Lowe), 46 (K. Gillett), 47 right (C. Arneson), 48 top (P. Human), 49 inset (I. Bennett), 52 (R.F. Myers), 53 top (David Masry), 53 bottom, 54 (US Naval Medical School), 55, 57, 59 bottom, 60 top (R.H. Knabenbauer), 60 lower (P. Saunders), 61 top left (K. Gillett), 61 top right (P. Saunders), 62 (K. Gillett), 63 top (N. Coleman), 64 left (R. Kreuzinger/Rosemary Watts), 64 right, 65 (K. Gillett), 66 left (A.B. Bowker), 71 top (R.C. Murphy), 72 top (P. Giocomelli), 72 bottom, 73 (H.L. Todd), 74 (US Naval Medical School), 76 top, 77 top (Smithsonian Institution), 78 top, 79 top, 81 top (L. Barlow), 83 bottom (US Naval Medical School, Halstead and Bunker), 84 top (US Naval Medical School), 84 bottom (H.L. Todd), 86 (US Naval Medical School/R. Kreuzinger), 88 bottom, 89 bottom left (R.H. Knabenbauer), 90, 91 top (R.F. Myers), 91 bottom (K. Gillett), 92 top (J.F. Myers), 92 bottom (J. Randall), 94, 103 top (D. Gotshall), 105 top (D. Gotshall), 106 top, 107 bottom, 108, 109 bottom, 110 top (R. Kreuzinger/Rosemary Watts), 110 bottom, 114 bottom (R. Myers), 115 top (R. Myers), 115 bottom (R. Kreuzinger/P. Mote), 116 bottom left (US Naval Medical School), 116 bottom right (R. Kreuzinger/Rosemary Watts), 117 top, 117 bottom (S. Calloway/Rosemary Watts), 118 top left (US Naval Hospital, R. Schoening), 118 bottom left, 118 right, 120 top (H. Heatwole), 120 bottom (R. Kruntz), 121 (H. Heatwole), 122 bottom, 123 top (L. Barlow), 123 bottom (H. Heatwole), 124, 125 top (US Naval Medical School), 125 bottom, 126 top (US Naval Medical School), 126 bottom, 127 top left, 127 bottom, 128, 129 top left (R. Kreuzinger/Rosemary Watts), 129 top right (B. Magrath), 130, 131, 133 top, 133 center right, 135, 139, 141 bottom, 142, 144 top (R.H. Knabenbauer), 146 top, 147 bottom left, 148, 149 bottom (R. Russo), 150 bottom left (R.F. Myers), 151 top left, 151 top right and bottom (R.F. Myers), 152 top and center, 153, 154 top (R.F. Myers), 156, 157 bottom, 158, 159 top, 160 center and bottom, 161 top (R.F. Myers), 161 bottom, 162 top, 162 bottom (US Naval Medical School), 163, 164, 165, 168 bottom, 180 bottom (P. Giacomelli), 181.

Planet Earth Pictures, London, England: pages 11 (N. Coleman), 12 top (F. Schulke), 13 (C. Petron), 14 top (C. Roessler), 15 top (N. Coleman), 15 bottom (A. Kerstitch), 16 bottom (R.H. Johnson), 19 (C. Roessler), 22 (M. Coleman), 23 bottom (C. Roessler), 24 (K. Lucas), 30 top (C. Petron), 30 bottom (K. Lucas), 31 top (H. Voigtmann), 13 bottom (C. Petron), 32 (C. Petron), 34 bottom (R. Salm), 35 top (L. Collier), 35 inset (C. Roessler), 37 top (B. Merdsoy), 37 bottom (N. Middleton), 38 (R. Salm), 39 (K. Lucas), 41 right (R.H. Chester), 42 (C. Roessler), 43 (P. Scoones), 44 (A. Mounter), 45 left (B. Wood), 45 right (W. Deas), 48 bottom (P. Atkinson), 49 top (L. Pitkin), 50 (D. Clarke), 51 (K. Vaughan), 52 top (D. George), 56 left (J. Lythgoe), 56 right, 58 (C. Petron), 59 top (C. Prior), 59 center (C. Petron), 61 bottom (C. Petron), 63 bottom (W. Deas), 66 right (C. Roessler), 67 (D. George), 68 (A. Double), 69 left (P.D. Capen), 69 right (A. Svoboda), 70 top (C. Petron), 70 center (J. Kenfield), 70 bottom (S. Weinberg), 71 bottom (W. Deas), 75 (D. Clarke), 76 bottom (W. Deas), 77 bottom (A. Svoboda), 78 bottom (B. Wood), 80 bottom (P.D. Capen), 81 bottom (A. Colclough), 83 top (K. Lucas), 85 (K. Lucas), 87 (C. Petron), 88 top (J. Greenfield), 89 top (P. Scoones), 89 bottom right (K. Cullimore), 93 top (K. Lucas), 93 bottom (P. Scoones), 95 (A. Svoboda), 96 top (C. Petron), 96 bottom (P. Scoones), 97 (K. Lucas), 98 top (N. Sefton), 98 bottom (K. Lucas), 99 (P.D. Capen), 100 (C. Petron), 101 top (C. Petron), 101 bottom (K. Lucas), 102 (D. George), 103 bottom (K. Lucas), 104 (K. Lucas), 105 bottom (K. Lucas), 106 bottom (C. Petron), 107 top (N. Coleman), 109 top (R. Manstan), 111 (K. Lucas), 112 top (K. Lucas), 112 bottom (C. Petron), 113 (R. Waller), 114 top (W. Deas), 116 top (J. Lythgoe), 119 (C. Roessler), 122 top left (C. Roessler), 122 top right (K. Lucas), 127 top right (J. Greenfield), 132 (P.D. Capen), 133 center left (D. Clarke), 133 bottom (B. Wood), 134 top (C. Petron), 134 bottom (R. Cehsher), 136-137 (D. Maitland), 138 (K. Lucas), 141 top (R.H. Johnson), 143 top (Ocean Images, W. Clayton), 143 bottom (C. Petron), 145 (K. Lucas), 146 bottom (C. Roessler), 147 top (C. Roessler), 147 bottom right (B. Wood), 149 top (H. Voigtmann), 150 top (B. Wood), 150 bottom right (L. Collier), 152 bottom (P. Scoones), 154 bottom (P.D. Capen), 155 (P. Scoones), 157 top (P.D. Capen), 159 bottom (C. Petron), 160 top (N. Sefton), 166 (R. Beales), 167, 168 top (C. Petron), 169 (B. Wood), 170 (D. Perring), 172 (A. Joyce), 173 (R. Salm), 179 (K. Lucas), 180 top (S. Weinberg).

Oxford Scientific Films, England: pages 171 (T.S. McCann), 178 (F. Ehrenstrom).

The color artworks on pages 21 (mackerel shark), 174 (sei whale), 175 (Southeast Asia porpoise), 176 (white whale), and 177 (sperm whale) were drawn by Graham Allen and are copyright © Marshall Editions Ltd, London. The color drawing of the European ratfish on page 82 is by Colin Newman and is copyright © Marshall Editions Ltd, London.

The illustration of the dusky shark (p.17) is from *FAO Species Catalogue Vol 4 Part 1, Sharks of the World*, ed. Leonardo V. Compagno (Rome, 1984); that of the cownose ray (p.79) is from *Poisonous and Venomous Animals of the World, Vol 3* by Bruce W. Halstead (Washington DC, 1970); and the illustration of the round stingray (p.80) is from *Marine Fishes of the Pacific Coast of Mexico*, ed. Tosio Kumada (Odawara, 1937), Thanks are due to Ralph Hutchings for photography and to the staff of the Zoology Department of the Natural History Museum, London, for their assistance.

Rosemary Watts drew the head and teeth illustrations in chapter 1.

Jackie Bond drew the distribution maps using information supplied by the World Life Research Institute.

INDEX

A

B

C

D

E

G

H

I

K

L

M

N

O

P

R

S

T

U

V

Z